A DICTIONARY OF CHROMATOGRAPHY

Second edition

A DICTIONARY OF CHROMATOGRAPHY

SECOND EDITION

R.C. DENNEY

A WILEY-INTERSCIENCE PUBLICATION

JOHN WILEY & SONS
New York

Published in the USA by

WILEY–INTERSCIENCE,

a Division of John Wiley & Sons, Inc.,
New York

ISBN 0 471-87477-9

Typeset by Leaper & Gard Ltd, Bristol
Printed in Great Britain

'Some books are to be tasted, others to be swallowed, and some few to be chewed and digested; that is, some books are to be read only in parts; others to be read but not curiously; and some few to be read wholly, and with diligence and attention. Some books also may be read by deputy, and extracts made of them by others.'

Francis Bacon
Essays, 50, 'Of Studies'

Introduction

During the past few years I have been greatly encouraged by the large number of people who have said how useful they have found the previous edition of *A Dictionary of Chromatography* and its companion volume *A Dictionary of Spectroscopy*. It is quite clear that in many ways I achieved the object I originally set of providing a source of reference for students, technicians and those newly interested in these particular areas of science.

In preparing this second edition, it has been apparent that there is still considerable conflict over the symbols used in the various areas of chromatography, despite attempts by IUPAC[1] to bring about a greater degree of standardization. The differences that remain have been admirably documented by L.S. Ettre[2,3] and should lead to further rationalization in the future. It is not the purpose of a volume such as this to arbitrate between the conflicting views. As far as possible I have used those symbols and equations which are in common use both in the USA and Europe and/or over which there is little likelihood of ambiguity. As far as is practicable I have standardized on S.I. units[4] but have retained the use of the Angstrom and commend the commonsense of the Royal Society[5] in accepting the use of L as the symbol for the litre as it is less likely to be confused with the letter l and the digit 1.

I would like to express my gratitude to several reviewers of the first edition who were kind enough to make constructive suggestions for improvements. Many of the new entries have been incorporated as a result of their comments and I am sure the book has benefited as a result. There is, however, always the problem of trying to establish a balance between what is apparently too trivial and entries which are totally remote and abstruse, at the same time I have wanted to provide a book which is of real value to as many people as possible. Because of this a number of the earlier entries have been increased in length, cross-referencing has been improved and the number of references increased by incorporating details of a large number of recent publications in expanding areas.

Introduction

Once again I would emphasize that this book, like *A Dictionary of Spectroscopy*, is not intended for the person who considers himself, or herself, an expert in the theory and practice of chromatography. No doubt they would have written the book differently and done it better! I would, however, be grateful for any constructive suggestions for inclusion of additional data and would appreciate readers drawing my attention to any errors that I have introduced, in the hope that further improvements may be possible in a future edition.

I would like to express my sincere thanks to those manufacturers and publishers who agreed to the use of their diagrams to help illustrate this volume and which have added to the overall quality of the presentation.

As usual I am greatly indebted to my long suffering family that continues to tolerate the noise of typing late at night, and to weekends devoted to trying to meet tight deadlines. This book, like previous ones, is as much a symbol of their tolerance as to my tenacity. I hope the final result achieves its purpose and justifies the efforts and demands placed upon a number of different people.

Ronald C. Denny

Symbols

The following symbols and letters are employed in the equations throughout the book. In some instances symbols which are only employed once, or with special meanings are explained in the text.

Å	Angstrom	D_s	surface activity distribution coefficient
A_i	peak area for solute i	D_V	volume distribution coefficient
A_R	detector response factor		
a	weight of dry gel	d_f	film thickness of stationary liquid phase
a_i	proportionality factor for solute i	d_p	average particle size
C_1	recorder sensitivity	d_R	retention distance on chromatogram
C_2	reciprocal chart speed		
C_3	mobile phase flow rate	E	cohesive energy
C_G	coefficient of mass transfer in gas phase	$\boldsymbol{E}$	electric field strength
C_s	coefficient of mass transfer in stationary liquid phase	E_u	lower limit of detectability
		F	nominal linear flow
c	solute concentration	F_c	volumetric flow rate corrected for column temperature
D	detector output		
D	electric displacement		
D_c	concentration distribution ratio	g	gram
		H	height equivalent to an effective plate
D_G	diffusion coefficient in gas mobile phase	h	height equivalent to a theoretical plate
D_g	distribution coefficient for gels	h_r	reduced plate height
		I	retention index
D_L	diffusion coefficient in stationary liquid phase	I	detector current
D_M	diffusion coefficient in mobile phase	j	column pressure gradient correction factor

Symbols

K_{av}	proportion of gel medium available to solute	r	radius of particle
K_c	concentration distribution ratio	$r_{i,s}$	relative retention
		r_o	radius of capillary
K_D	distribution constant (or partition coefficient)	S	sensitivity
		S	separation factor
k	kilo-	S_r	solvent regain
k	mass distribution ratio (capacity factor, capacity ratio or partition ratio)	T	temperature
		t	time
		t_M	retention time for non-sorbed species, hold-up time
L	litre		
L	column length	t_R	retention time
M	migration value	t_R^0	corrected retention time
M_r	molecular weight	t_R'	adjusted retention time
m	milli-	u	interstitial velocity
n	nano-	$\bar{u}$	average linear gas velocity in column
n	number of theoretical plates		
N	effective theoretical plate number	u_E	electrophoretic mobility
		u_i	inverse speed of chart paper
p	pressure	V	volt
p_i	column inlet pressure	V	volume
p_0	column outlet pressure	V_d	dead volume
Q	quantity of sample	V_e	elution volume
Q	molecular ionization constant	V_G	interstitial volume for GC
		V_g	specific retention volume
Q_E	net charge on particle	V_I	inner or interstitial volume
R	gas constant		
R_b	retention factor relative to a standard	V_i	stationary liquid volume
		V^1	molar volume of liquid
R_F	retention factor (or relative front)	V_L	liquid stationary phase volume
R_M	relative movement for related compounds	V_M	hold-up volume, gas hold-up volume
R_s	resolution	V_N	net retention volume
R_u	background noise	V_o	void (outer) volume

V_R	total retention volume for solute	δ	solubility parameter
V_R'	adjusted retention volume	ϵ_i	interstitial fraction
V_R^0	corrected retention volume	ϵ_r	dielectric constant
V_S	volume of solid packing	ϵ_s	stationary phase fraction
V_t	total volume	ϵ^0	solvent strength
w_b	peak width at base	ζ	zeta potential
w_h	width at half peak height	η	viscosity
w_i	weight of sample i	$[\eta]$	limiting viscosity number
w_L	weight of stationary liquid phase in column	λ	packing constant
w_r	water regain	μ	statistical mean
X	column volume	μ	micro-
X	detector signal	ν	reduced velocity
x	molar fraction of solute	π	pi
α	separation factor	ρ	specific gravity of solvent
α^*	retention volume ratio	σ	standard deviation
β	phase ratio	σ_d	detector sensitivity
γ	labyrinth factor	σ_e	sensitivity of chart recorder
		Φ	Flory-Fox constant

Greek Alphabet

As so many equations include the use of Greek letters the following alphabet has been included to help the reader to give the correct name to these letters as they arise in the text:

alpha	A	α	nu	N	ν
beta	B	β	xi	Ξ	ξ
gamma	Γ	γ	omicron	O	o
delta	Δ	δ	pi	Π	π
epsilon	E	ϵ	rho	P	ρ
zeta	Z	ζ	sigma	Σ	σ
eta	H	η	tau	T	τ
theta	Θ	θ	upsilon	Υ	υ
iota	I	ι	phi	Φ	ϕ
kappa	K	κ	chi	X	χ
lambda	Λ	λ	psi	Ψ	ψ
mu	M	μ	omega	Ω	ω

Abbreviations

The following abbreviations are used throughout this book:

GC	gas chromatography
GLC	gas liquid chromatography
GSC	gas solid chromatography
HPLC	high performance liquid chromatography
LC	liquid chromatography
PC	paper chromatography
TLC	thin layer chromatography

Sources of Diagrams

The following figures have been reproduced with the permission of the copyright holders from the sources listed: Fig.1, D.R. Browning (ed.), *Chromatography* (McGraw-Hill, London and New York, 1969), Figs. 4 and 45, D.F.G. Pusey, *Chem. Brit.*, **5**, 408 (1969), Fig. 5, R.C. Denney, *A Language of Its Own – Chemistry* (Muller, 1982); Fig. 2, F.W. Fifield and D. Kealey, *Principles and Practice Of Analytical Chemistry* (International Textbook Co. Ltd, 1975); Fig. 10, Farrand Optical Co. Inc., Mount Vernon, New York; Fig. 11, E.L. Durrum, *J. Amer. Chem. Soc.*, **73**, 4875 (1951), Fig. 13, Finnigan Instruments Ltd, Hemel Hampstead, Hertfordshire; Fig. 25, Griffin and George Ltd, 285 Ealing Road, Wembley; Figs. 27 and 39, Perkin Elmer Ltd, Beaconsfield, Buckinghamshire; Fig. 41, A.C. Stern (ed.), *Air Pollution*, vol. 2 (Academic Press, London, 1968); Figs. 48 and 55, S.G. Perry, *Chem. Brit.*, **7**, 366 (1971); Fig. 49, Pharmacia Fine Chemicals AB, Uppsala, Sweden. All other illustrations are the copyright of the author.

A

Absolute detector sensitivity. The change required in the physical parameter that will produce a full-scale deflection of the recorder at maximum detector sensitivity for a defined level of **noise**. It is employed particularly in connection with HPLC detectors. *See also* **Relative sample sensitivity**; **Sensitivity**; **Signal-to-noise ratio**.

Absolute retention volume. *See* **Net retention volume**.

Absorption. Penetration in bulk of one material throughout a second material. It is usually associated with the formation of fairly strong chemical bonds and is rarely reversible at ambient temperatures. Dissolution of gases in liquids frequently involves absorption forces. The main difference between this and **adsorption** is that the former involves fairly uniform penetration of the absorbent matrix, while the latter occurs predominantly in surface layers, and the forces involved in adsorption are very much less than the chemical forces of the absorption process.

Acidic cation exchange. *See* **Cation exchanger**.

Activation of adsorbents. In general the activity of the **adsorbents** employed in TLC and other forms of adsorption chromatography is increased as the water content is reduced. To obtain the greatest degree of activity, **kieselguhr** and **silica gel** plates are usually dried at 100 °C for one hour. They are then stored in a dry chest until required as the activity is lost by adsorption of atmospheric moisture. **Alumina** plates vary greatly in activity depending upon the period of time and temperature at which drying has been carried out. The activity of alumina columns and plates is established by use of the **Brockmann scale** of dyes.

Adjusted retention time (t'_R). *See* **Retention time**.

1

Adjusted retention volume (V'_R). The **retention volume** for the substance less the **hold-up volume** (the retention volume for a nonsorbed species):

$$V'_R = V_R - V_M = t'_R F_c$$

Adsorbents. Any material which will adsorb one substance in preference to another from solution can be employed as an adsorbent in adsorption chromatography. As a result a very wide range of materials has been used. Those most commonly employed for GSC include activated carbon, **silica gel** and microporous polymers. For TLC the most common are **alumina, cellulose, kieselguhr** and silica gel. In some cases the adsorbents for TLC contain a small amount of added gypsum which acts as a binder (for example in Silica gel G). To assist detection of zones they may also contain a fluorescing indicator.

Adsorption. A process which occurs at the surface of a liquid or solid as a result of the attractive forces between the adsorbent and the solute. These forces may be physical, such as **van der Waals' forces**, or weakly chemical, as in the case of **hydrogen bonding**. Physical adsorption is associated with low heats of adsorption whilst chemical adsorption usually involves higher energy changes. As a result chemical adsorption is stronger than physical adsorption.

Adsorption probably occurs at all surfaces and interfaces to some extent, but for use in separative processes it is common to employ porous substances possessing large effective surface area to mass ratios.

Adsorption processes such as those in GSC are normally carried out at higher temperatures than are the corresponding separations by partition on GLC. The higher temperatures are necessary in order to obtain separations in a reasonable period of time from the GSC column.

Adsorption chromatography. A form of chromatography which employs an adsorbent such as **silica gel** or **alumina** as a solid stationary phase, and either a liquid mobile phase (in TLC and adsorption column chromatography) or a gas mobile phase (in GSC). Separation is dependent upon the different extents to which solutes are adsorbed by the solid: the less strongly adsorbed materials travel faster than the strongly adsorbed materials.

The introduction of adsorption chromatography is usually credited to Michel Tswett,[6,7] although his work on the separation of plant pigments was published at about the same time as Day's report[8] on the separation of coloured oils by percolation through earth. Real credit for the first useful studies on adsorption columns should, however, go to Reed[9] whose work, published in 1893, included separations both of iron(III) chloride and copper(II) sulphate, and of alkaloids on columns of powdered kaolin.

Adsorption detector. *See* **Thermal adsorption detector.**

Adsorption isotherm. In adsorption chromatography the **distribution coefficient** is dependent upon the concentration of the substance in solution. At a constant temperature the distribution of the solute between the solid adsorbent stationary phase and the liquid mobile phase can be represented in a graphical form as the adsorption isotherm, being a pictorial presentation of the concentration in the mobile phase (per unit volume) in equilibrium with the corresponding concentration in the stationary phase (per unit weight). The shape of the adsorption isotherm is a guide to the performance of the solute and adsorbent under TLC conditions, as shown in Figure 1. The ideal linear isotherm, representing constant distribution between the two phases at all concentrations, is rarely achieved over a large concentration range as all substances in solution affect each other. The convex isotherm (Figure 1b) gives rise to tailing bands or spots, while the concave isotherm (Figure 1c) produces leading bands or spots. It is tailing of this type which frequently makes resolution in adsorption chromatography so difficult. Figure 1d and 1e indicates how the angle of the line is related to the R_F **value** for the solute.

With **zeolites** the adsorption isotherm is a measure of the amount of material adsorbed related to the applied pressure.

Aerogels. If removal of the dispersing agent from a gel system can be achieved without shrinkage of the gel structure the rigid form obtained

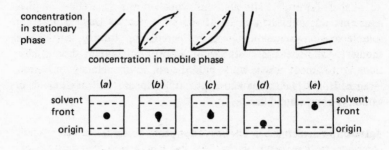

concentration in stationary phase

concentration in mobile phase

(a)   (b)   (c)   (d)   (e)

solvent front

origin

solvent front

origin

Figure 1. Relationship between adsorption isotherm and sample movement on thin layer chromatography

is called an aerogel. In practice this name is now applied to any rigid structure that can be employed for **gel chromatography** even if it is not strictly a gel. The term not only includes silica gel beads but also the modern glass aerogels[10] prepared by heat treating borosilicate glass to produce channels with regular dimensions of 75-2000 Å (7.5-200 nm) in the glass powder. These rigid glass and silica gel aerogels are of particular importance for gel work carried out by HPLC as they retain their structures even under pressure, while **xerogels** will collapse.

Affinity chromatography. A procedure first described by Campbell *et al.*[11] in 1951 and used to obtain a highly selective purification of biomolecules by employing a medium consisting of a gel matrix combined with an enzyme or nucleic acid.[12] The biospecific adsorbent is prepared by coupling a specific ligand (such as an enzyme, antigen or hormone), for the macromolecule of interest, to a water-insoluble carrier by means of a free carboxyl, amino or phenolic group.[13] By making the column selective or specific in this way only those molecules able to bind with the immobilized ligand are retarded and held on the column,[14,15] to be later released in a purified state as required.[16] The range of immobilized ligands is very large and extends from simple molecules such as boronic acid[17] to deoxycholic acid.[18]

Agar-agar. A polysaccharide obtained by extracting red seaweed; it is soluble in hot water and on cooling the solution forms a firm gel. Although agar itself has been employed for chromatographic separations,[19] it is more commonly separated into its two main components — **agarose**, the neutral component, and agaropectin, which contains carboxylic and sulphonic acid groups.[20]

Agarose. The neutral component of agar-agar. It is obtained after acidic components have been precipitated by acetyl pyridinium chloride[21] or by fractional precipitation with polyethylene glycol.[22] For **gel chromatography** the Agarose has to be in a special purified form of homogeneous beads.[23]

Chemically it is a high molecular weight polysaccharide[24] formed from β-D-galactopyranose and 3,6-anhydro-α-L-galactopyranose units, with a composition corresponding to $[C_{12}H_{14}O_5(OH)_4]_n$. The gel is used for separating substances with high molecular weights in the range 10^3-10^8 and for studies on viruses, phages and bacteria. Only aqueous solutions can be employed as organic solvents destroy the gel structure.[25] It is also used in a form in which it is cross-linked with dextran and reacted with compounds such as carbonyldiimidazole to give an activated matrix suitable for **affinity chromatography**.

Alkali metal flame ionization detector. *See* **Thermionic detector**.

Alumina. Chemically represented as Al_2O_2, xH_2O, one of the most common materials employed as a **stationary phase** in adsorption chromatography. It has a large surface area and a high porosity providing a retentive, basic surface suitable for separating neutral and basic compounds. It acts as an adsorbent by forming **hydrogen bonds** between its own hydroxyl groups and functional groups on the solute. It is frequently used with no other additives, but may also contain 5% gypsum as a binder (Alumina G), and/or fluorescent indicators for viewing spots under ultraviolet light. The activity of alumina, which is determined by the **Brockmann scale**, decreases with its water content. It is usually available in five grades, the most active being grade I.

Alumina G. *See* **Alumina**.

Ambient operation. A term used most commonly in connection with HPLC, referring to operation of the chromatograph at room temperature without the application of **temperature programming** or elevated temperature. **Flow programming** and **gradient elution** are usually carried out under ambient conditions. *See also* **Subambient operation**.

Amino acid analyser. A very specialized application of **ion exchange chromatography** by which amino acids are differentially retarded, separated and eluted in a regular sequence according to their basic, neutral and acidic characteristics. The process, first described by Spackman *et al.*[26] in 1958, may employ as many as five ion exchange columns and either a series of buffer solutions or a gradient buffer mixture.[27] The detection of individual amino acids in the eluate is carried out photometrically[28] by the measurement of absorbances at 440 nm and 570 nm from colours produced between the amino acids and ninhydrin reagent at 100 °C. Amino acid analysers are manufactured commercially and compact systems have been designed.[29,30] Recent advances have led to the development of single column systems capable of accepting sample volumes of up to 30 μL with detection limits below nanomole concentrations.

Ampholines. A trade name for a series of carrier **ampholytes** used in **isoelectric focusing** and based upon a polyaminopolysulphonic acid gel structure.

Ampholytes. Substances which carry both positive and negative charges (i.e. they are amphoteric). Typical ampholytes are amino acids and proteins which possess both $-COO^-H^+$ and $-NH_3{}^+OH^-$ groups and for which the structure at any time is pH dependent in the following manner:

$$H_2N-CH-COO^- \rightleftharpoons H_3N^+-CH-COO^- \rightleftharpoons H_3N^+-CH-COOH$$

R	R	R
at alkaline pH	at isoelectric point	at acid pH

They are used particularly in **isoelectric focusing** for forming a pH gradient.

Amphoteric ion exchange resins. Special bipolar ion exchange resins carrying both positive and negative ionic active groups. Both types of active group are able to participate in ion exchange processes, frequently carried out simultaneously on the same solution. Studies have shown[31] that the single groups react independently and stoichiometrically.

They consist of a linear cross-linked mixture of polymers comprising a cation exchanger entwined within an anion exchanger. This structure is achieved by polymerizing the cation exchanger monomer within the structure of the previously formed anion exchanger producing the snake cage structure required for **ion retardation**. Such resins are employed to remove electrolytic impurities from liquids[32] by the preferential retardation of the electrolyte due to the formation of ion pairs with the resin.

Anion exchanger. Any **ion exchanger** solely capable of exchanging anions. The synthetic ion exchange resins used for this purpose are insoluble polymeric bases which produce insoluble salts as a result of the ion exchange operation. There are two main categories of anion exchanger – the strong base exchangers and the weak base exchangers. A typical strong anion exchanger is prepared from the polymerization of styrene and divinylbenzene, followed by chloromethylation and treatment with an appropriate secondary or tertiary amine (see equation on previous page).

The weak anion exchanges contain a mixture of primary, secondary and tertiary amino groups. They are produced by condensing aliphatic amines with either epichlorohydrin or ethylene dichloride:

$$> NH + CH_2-CH_2-CH_2\,Cl + RNH_2 \longrightarrow\ > N-CH_2-CH-CH_2-\overset{+}{N}R$$
$$\underset{O}{\diagdown\ \diagup} \qquad\qquad\qquad\qquad\qquad \underset{OH}{|}$$

Similar weak base exchangers are prepared by treating the chloromethy-lated styrene–divinylbenzene system with an appropriate secondary amine. Other anion exchangers are made by substituting diethylamino-ethyl groups into a cellulose matrix, the resulting exchangers being weakly basic suitable for operation within a pH range of 2-9.

Weak anion exchangers are not ionized at high enough pH to remove weak acids, and only the strong anion exchangers containing quaternary ammonium groups are suitable for this purpose. *See also* **Deionized water**; **Mixed-bed column**; **Cation exchanger**.

Anode. The positively charged electrode in any electrical circuit to which negatively charged particles and ions are attracted. For zone **electrophoresis** anodes are made from graphite, platinum or silver/silver chloride, those of platinum being preferable because of their robustness and chemical stability. In most electrophoresis cells the system is reversible so that anode and **cathode** are identical and interchangeable.

Apiezon L. Separation of hydrocarbons and other substances with low polarity can be readily achieved by GLC by employing Apiezon L as the stationary phase. It is a long-chain hydrocarbon of general formula $C_n H_{2n+2}$, used up to a 10% loading on the support with a maximum operating temperature of about 250 °C. Two similar substances are Apiezon M and Apiezon N.

Applicator. Any of a variety of means by which samples for PC and TLC are applied to the sheet or layer. The simplest of these is a small wire loop about 1 mm diameter dipped in the sample solution, then touched to the surface. For semiquantitative work a higher degree of accuracy is obtained by using **microcapillaries** or **micropipettes** capable of dispensing quantities up to 100 μL.

The special applicators used for applying large volumes of solution as streaks for thick layer preparative chromatography consist of a hypodermic syringe and wide bore needle attached to a horizontally moving carriage fixed above the adsorbent layer.

Argon detector. A detector, first described by Lovelock[33,34] in 1958, which has the advantage of being simple to construct and of having a high sensitivity suitable for trace analysis and qualitative work. Its main disadvantage is that it is easily overloaded, but its lower limit of detectability is lower than that of the **flame ionization detector**.

Operation of the detector (Figure 2) is based upon the excitation of argon carrier gas to a metastable state by collision with accelerated secondary electrons produced from the ionization of other argon atoms. The source of the primary radiation causing the ionization of the argon is a radioactive β-emitting isotope such as ^{90}Sr or ^{63}Ni. Any solute entering the detector with the carrier gas and possessing an ionization energy less than the energy of the metastable argon suffers an energy transfer resulting in the ionization of the solute molecules. This produces an increase in current across the detector which is amplified and recorded. The response of the detector is quenched by any water in the gas flow and it does not respond to hydrogen, nitrogen, oxygen, carbon dioxide, carbon monoxide, methane, halogens or fluorocarbons. For other substances it has a sensitivity of 10^{-10}–10^{-13} g cm^{-3} in the carrier gas.

Microargon detectors have been designed,[35] and commercial versions have an effective volume of less than 1 cm^3.

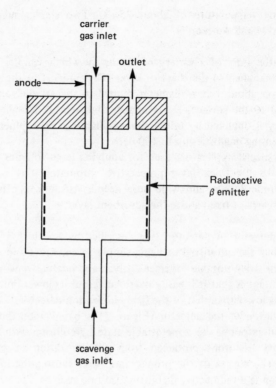

Figure 2. Argon detector

Ascending chromatogram. In both PC and TLC the running of chromatograms may be carried out by the solvent ascending the sheet or plate, which is maintained in a vertical or near vertical position with the bottom edge immersed in a solvent about 1 cm deep (see Figure 5, page 28). Movement of the solvent depends upon capillary action and the rate of movement of the solvent front decreases with time. As a result, ascending chromatograms require a greater time to obtain a good separation than do **descending chromatograms**, but in the case of TLC are much easier to carry out successfully.

Atomizer. The apparatus used to produce the fine mist of a reagent which is sprayed onto paper and thin layer chromatograms. The pressure needed to generate the spray may come from either a hand-operated rubber bulb or an aerosol spray can. The term atomizer is used in a different sense in atomic absorption spectroscopy.

Autoradiography. One of the special techniques applied to the detection of chemial compounds labelled with radionuclides, and particularly suited for PC and TLC. For this purpose, a specially sensitive photographic film is carefully placed on the surface of the chromatogram for a predetermined period of time. The presence of the radioactive compounds is shown on the film by black exposed spots and for semiquantitative purposes the intensity is roughly proportional to the concentration of the active material.[36]

Exposure time required is dependent upon a number of variables, including film sensitivity, type of radionuclide and the nature of the emitted radiation.

Available capacity. The actual capacity of the **ion exchange resin** that is available for exchange under defined experimental conditions. The available capacity is always less than the total capacity of the resin as some ion exchange sites may be inaccessible to the ions being exchanged. Temperature, pH and the nature of the eluent all affect the available capacity. It also varies slightly between different ions and their abilities to penetrate the resin. Resin capacities are expressed in terms of milliequivalents per gram of the wet resin. *See also* **Exchange capacity**.

Axial eddy diffusion. The former name for what was originally the second term of the **van Deemter equation**.[37] It is now more common to present this as the first term of the equation, as it is the only term not involving the linear mobile phase velocity, and to call it the **eddy diffusion term**.

Axial molecular diffusion. The former name for what was originally the first term in the **van Deemter equation**.[38] It is now referred to as the **molecular diffusion term** and is placed second after the **eddy diffusion term**.

B

Backflushing (Carrier gas inversion). The technique of reversing the flow of carrier gas in gas chromatographic columns. Its main purpose has been[39] for flushing out slow moving components from the column before they have had the opportunity to travel far along the column. It is only of advantage if the substance of interest moves relatively much faster such that there is a gain of instrument time by operating the backflushing technique. It has also been used to clear columns which have become partially blocked at the injection end of the column.

Back-pressure device. As pressure in an HPLC system varies from the applied inlet pressure to atmospheric pressure in the detector there is the possibility that gases (usually air) dissolved in the solvent will be released from solution inside the detector, upsetting the signal and giving a misleading reading. To obviate this the back-pressure device is employed to apply a back pressure of up to 20 pounds per square inch ($\approx 14 \times 10^4$ Nm^{-2}) to the solvent as it passes through the detector. This is achieved by fitting an adjustable valve and pressure gauge to the flowline after the detector. Partial closing of the valve enables an appropriate pressure to be attained.

Band broadening. A process which occurs in all forms of chromatography whereby the width of the zone slowly increases the longer the chromatogram is run. The occurrence of broadening is an indication that a true equilibrium between the amount of material in the mobile phase and that in the stationary phase is never really attained. The three factors influencing band broadening are those corresponding to the three terms of the **van Deemter equation**[40] — the **eddy diffusion, mass transfer** and **molecular diffusion terms**. At low flow rates of mobile phase the eddy and molecular diffusion terms are the main contributing factors to band broadening, but at high flow rates it is the mass transfer term that is mainly responsible. In the latter case the band broadening is a direct measure of the rate of flow relative to the rate of mass transfer.

Baseline. When only the mobile phase is passing through the detector with no solute in solution, the signal from the detector to the recorder should be a constant one and corresponds to what are termed baseline conditions, producing the steady baseline portion of the chromatogram from which all quantitative measures are made. *See also* **Integration** and **Triangulation**.

Baseline drift. Any regular change occurring in the baseline signal from the detector in GC or HPLC. It commonly arises from changes in the flow rate of the mobile phase or from deterioration of the detector such as might occur if the stationary phase is **bleeding** off the column.

Basic anion exchanger. *See* **Anion exchanger**.

Batch process. The most common and most easily applied procedure employed for preparative or continuous chromatography. It really consists of a scaled-up version of the normal analytical procedures: constant large volumes of samples are regularly applied to chromatographic columns constructed with larger internal diameters than for conventional analytical columns. The process is very tedious to carry out manually so that automatic injection of the sample from a bulk volume is usually made, and the collection of required fractions is by an automatic **fraction collector**. By this means up to kilogram amounts per hour can be separated. The disadvantage of the process is that there are long periods during which no sample can be injected as further injections are only possible when all components have been eluted from the column. Also it can take a very long time to fully separate large quantities of material unless there is a substantial difference in properties between the required substances and the impurities. *See also* **Circular columns**; **Moving bed process**; **Radial flow columns**.

Bed volume. *See* **Column volume**.

Bed volume capacity. *See* **Exchange capacity**.

Beilstein flame detector. Essentially a **flame ionization detector** fitted with either a brass head or a copper spiral in the tip of the flame.[41,42] It is intended as a selective detector suitable for the determination of halogens, based upon the traditional Beilstein test in which the flame becomes coloured green due to the formation of the copper halide. The main disadvantage of the detector it that it is not automatic and the operator has to observe the coloured flame and mark the appropriate peaks on the chromatogram. It can, however, detect 0.2 μg s^{-1} of halogen.

Belt transport detector. A variation on the **transport detector** which has been applied with particular success to **gel chromatography** in HPLC systems.[43] Instead of using a wire to collect a portion of eluate from the column, it employs a metal ribbon forming an endless belt between two pulleys to collect *all* the effluent.[44] As in the conventional transport detector, the solvent is evaporated and the solute pyrolysed. The products of the pyrolysis are carried to a flame ionization detector by a stream of helium and detected in the usual manner. It has been employed for separative studies on polymers. *See also* **Disc conveyor flame ionization detector**.

Binders. Substances frequently added to the adsorbents employed in TLC to assist their adhesion to the glass or plastic sheet. Gypsum, polyvinyl alcohol and starch have all been used for this purpose. Adsorbents containing 5% gypsum as a binder are sold commercially as Silica gel G and Alumina G. The presence of the binder also helps to prevent cracking of the layers once the plate has been prepared and dried.

Bioautography. The only general method of detection in chromatography that is available for antibiotics. It was first introduced[45] in 1946. In this procedure the paper or thin layer chromatogram is pressed for up to 30 minutes onto a nutrient layer containing specially selected micro-organisms. During this time some of the antibiotic diffuses into the nutrient in the locations corresponding to the spots on the chromatogram. On incubation of the plates the growth of micro-organisms is inhibited at the zones occupied by the antibiotics which have come from the chromatogram.[46] These positions are then measured with respect to the original paper or thin layer. The most frequently used micro-organisms for this purpose are *Bacillus subtilis*, *Escherichia coli* and *Staphylococcus aureus*.

Bleeding. Liquid stationary phases employed in GLC have maximum operating temperatures. When used below the particular operating limit, loss of the stationary phase from the solid support is negligible, but if the column is heated above the limit there is a progressive loss, or bleeding, of the stationary phase from the column in the carrier gas flow. The operating limit for **squalane**, for example, is 160 °C and that for **dinonyl phthalate** is 150 °C. Apart from deterioration in the separative properties of the column, bleeding leads to **baseline drift** due to the stationary phase passing through the detector, and in the case of silicone oils can lead to silica being deposited on the detector surfaces.

Blue dextran. Determination of the **interstitial** (void) **volume** on gel columns is carried out by employing high molecular weight substances which are totally excluded by the gel. The most widely used material for this purpose is Blue dextran. Blue dextran 2000 is prepared from dextran possessing an average molecular weight of 2×10^6, by the incorporation of a small amount of a polycyclic chromophore. It is employed as a 0.2% solution in phosphate buffer at pH 7.06. It is not suitable for **thin layer gel chromatography** and for this purpose **Orange dextran** has been developed.[47]

Bonded stationary phases. A wide variety of column packing materials in which the active stationary phase is a substance which is chemically bonded to an inactive, essentially inert support material. Most commonly[48] the support material consists of silica or controlled-pore glass beads, and typical bondings involve interaction between hydroxyl groups on these surfaces with a variety of silanes possessing various functional groups ranging from $NH_2 CH_2-$ to substituted aromatic systems. Such bonded stationary phases are used extensively in **reversed phase systems**. **Pellicular coatings** are a special type of bonded stationary phase in which a thin layer of ion exchanger is held firmly on the surface of a glass microbead.[49]

Breakthrough capacity. *See* **Exchange capacity**.

Brockmann scale. In order to assist the classification of alumina for adsorption chromatography according to the various grades of activity, Brockmann and Schodder[50] proposed the use of a series of dyes for which the adsorbability increased in the order

azobenzene (ab) < p-methoxyazobenzene (mab)
    < sudan yellow (sy) < sudan red (sr)
        < p-aminoazobenzene (aab) < p-hydroxyazobenzene (hab)

The dyes are usually used in pairs and the activity of the alumina classified according to which dye is adsorbed at the top of the column while the adjacent one on the list is being eluted, as shown in Table 1.

Table 1

	Alumina activity				
	I	II	III	IV	V
Dye at column head	mab	sy	sr	aab	hab
Dye at column bottom	ab	mab mab	sy sy	sr sr	aab
Eluted		ab	mab	sy	

The activity of the alumina can also be determined from the R_F values obtained for these dyes,[51] and these are listed in Table 2. The most suitable solvent for development with these azodyes for the above determinations was found to be carbon tetrachloride.

Table 2

Dye	Alumina activity			
	II	III	IV	V
Azobenzene	0.59	0.74	0.85	0.95
p-Methoxyazobenzene	0.16	0.49	0.69	0.89
Sudan yellow	0.01	0.25	0.57	0.78
Sudan red	0.00	0.10	0.33	0.56
p-Aminoazobenzene	0.00	0.03	0.08	0.19

Brunel mass detector. *See* **Mass detector.**

Buffer solutions. Solutions capable of resisting changes in the hydrogen ion concentration when diluted or when small quantities of acids or bases are added to them. Buffer solutions usually consist of a mixture of a weak acid and the sodium or potassium salt of that acid, or of a weak base and its salt — the buffer is, therefore, a mixture of an acid and its conjugate base.

In chromatography, buffer solutions are of importance for gel permeation and ion exchange studies. For electrophoresis, buffer solutions are essential for two reasons: first to act as a conductor for the current, and second to maintain constant pH conditions during the separation. A wide range of buffer solutions is used for these purposes; the most common are ammonia with ammonium acetate, pyridine with glacial acetic acid, sodium barbitone with sodium acetate, and EDTA with boric acid.

Bulk property detector. Any detector which measures a change in a physical property of the mobile phase. The term is used most frequently in connection with the detectors in HPLC. For example, conductivity, dielectric constant and refractive index detectors all come within this classification as they all measure changes in the initial property of the solvent resulting from the presence of the solute. Such detectors are usually less sensitive than **solute property detectors** such as the **ultraviolet absorption detector**. They tend to suffer from noise and drift due to difficulties of maintaining the rigorous isothermal conditions that are required.

By-pass injector. A type of injection system used for injecting gaseous samples in GC. It involves a sample chamber, of predetermined capacity, that is shut off from the main carrier gas stream by taps or valves at each end of the chamber. After the chamber has been filled with the sample it is swept into the column head, with virtually no dilution, by diverting the carrier gas into the by-pass chamber on turning the taps, as seen in Figure 3. The operation of the device differs from that of **sample loops** in which it is the sample which is rotated into the gas stream.

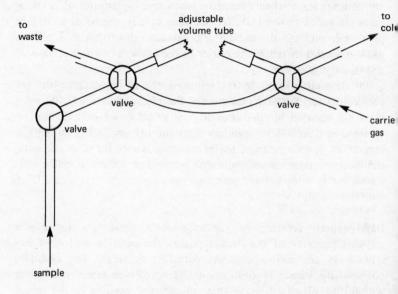

Figure 3. By-pass injector

C

Calibrated leak device. *See* **Sample splitter**.

Capacity factor (Capacity ratio). *See* **Mass distribution ratio**.

Capacity of ion exchange resin. *See* **Exchange capacity**.

Capillary columns (Golay columns). Although the possibility of using capillary columns was first suggested by Martin,[52] it was Golay[53,54] who carried out the first practical separations employing capillary columns in which the stationary liquid phase had been coated on the inside walls. It was found that these columns had the advantages of low resistance to gas flow and high efficiencies of thousands of plates.

Capillary columns have been made from copper, glass, stainless steel and even nylon; they have been up to 300 metres long and with internal diameters of 100-500 μm. As only one gas path is possible through these columns the eddy diffusion term of the **van Deemter equation** no longer applies and a modified equation,[55] frequently known as the **Golay equation**, is used for the determination of the height equivalent to a theoretical plate.

The original capillary columns are classified as **WCOT columns**, and since their introduction various other forms of treated capillary columns have been developed. *See also* **PLOT columns** and **SCOT columns**.

Capillary flow. Evaporation of the **buffer solution** occurs in zone **electrophoresis** as a result of the current heating effect. This evaporation causes replacement of the solvent by capillary flow from the ends of the strip towards the centre, and it retards the electrophoretic migration. It is best avoided by efficient cooling to remove the heat generated, and/or by operating with **enclosed strip electrophoresis**.

Capsule sampling. A method of applying fixed volume liquid and solid samples to GC. The samples are sealed in a 20 μL aluminium capsule

19

which is cold welded. The capsule is then mounted in a probe and inserted into a GC injection block through a pressure-tight lock system where a fixed needle pierces the capsule and the sample is transported by carrier gas to the column.[56] This sampling procedure has the advantage of providing solvent-free analyses, and nonvolatile components remain in the capsule without clogging the column head or the inlet system.

Carbowax. Any of one of the most widely used groups of liquid **stationary phases** for GLC. Carbowax is a trade name for polyethylene glycols or methoxypolyethylene glycols possessing the general formula

$$RO-CH_2-CH_2-(O-CH_2-CH_2)_n-OR \text{ where } R = H \text{ or } -CH_3$$

The number of carbowax is an indication of the average molecular weight; for example Carbowax 600 is a polyethylene glycol with a molecular weight range of 570–630.

As the Carbowaxes are themselves polar, they are suitable for separating polar substances such as alcohols. Operating limits vary from about 150 °C for Carbowax 1500 to over 200 °C for the high molecular weight Carbowaxes.

Carrier. Used particularly in **affinity chromatography** to refer to the water-insoluble polymer used to carry the active ligand. The carrier should form a chemically and mechanically stable medium without interfering with the essential properties of the ligand (enzyme, hormone, etc.).[57] The most common method of using the carrier is by covalent coupling between it and the ligand, as is done with **Sephadex. Agarose** and with cellulose derivatives. Porous glass[58,59] and various polymers have also been used for this purpose. Enzymes, for example, can be readily bound to polyacrylamide gels, and antigens to specially activated forms of Agarose.

In liquid-liquid partition TLC, the solid sorbent of the stationary liquid phase is termed the carrier.

Carrier gas. In gas chromatography, the mobile phase of the separative system. The most commonly used gases for this purpose are hydrogen, helium and nitrogen, although argon, carbon dioxide, air and neon are used under particular conditions. Often the type of **detector** employed dictates the nature of the carrier gas to be used, for example hydrogen or helium for a **katharometer** and argon or nitrogen for the **electron capture detector**.

The gases are supplied at a constant pressure and flow rate from high pressure cylinders and frequently require further purification and drying by passage through **molecular sieves** and drying agents before being suitable for use on the gas chromatograph. Hydrogen has been shown[60] to give greatest column efficiency at operations above 70 °C and nitrogen below 70 °C.

Flow rates vary greatly, but are normally between 20 and 100 cm^3 min^{-1} for normal GC. Optimum flow rates can be calculated from the **van Deemter equation**. For preparative GC other factors come into consideration and higher flow rates of 200–750 cm^3 min^{-1} are used.

Carrier gas inversion. *See* **Backflushing.**

Catharometer. *See* **Katharometer.**

Cathode. The negatively charged electrode in an electrical cell to which positively charged particles and ions are attracted. In zone **electrophoresis** the current in the circuit is carried by the ions of the electrolyte, and at the same time substances with a net positive charge on the electrophoretic strip move in the direction of the cathode. Like the **anode** this electrode is most commonly made from graphite, platinum or silver/silver chloride. It is interchangeable with the anode.

Cation exchanger. An **ion exchange resin** capable of exclusively exchanging cations. The early cation exchangers were naturally occurring **zeolites** but these have been almost completely superseded by the synthetic organic ion exchangers.

The synthetic cation exchangers are classified into two main groups, the weak cation exchangers and the strong cation exchangers. In the former the main functional groups are carboxylic acid groups and in the latter they are sulphonic acid groups.[61,62] The strong cation, or acid, resins are made either by sulphonation of phenol-formaldehyde polymers, sulphonation of styrene-divinylbenzene polymers, or by polymerization of phenolsulphonic acid with formaldehyde. The overall process with styrene and divinylbenzene gives an idea of the polymeric structure obtained:

The exchange capacity of strong cation exchangers is virtually independent of pH.

Weak acid exchangers are prepared from the polymerization of esters of unsaturated carboxylic acids (for example maleic or methacrylic acids) with styrene and/or divinylbenzene:

Weak cation exchangers only ionize in alkaline solution and are not suitable for use in solutions below pH 7. An alternative series of cation exchangers has also been created by substituting carboxymethyl groups into a cellulose matrix. These are also weakly acidic and may be used over a pH range of 3-10. *See also* **Anion exchanger**.

Celite. A **diatomaceous earth**, similar to **kieselguhr**, that is one of the many materials employed as solid **supports** in GC. It is an aggregate formed by fusing the diatomite with a small amount of sodium carbonate flux at 900°C. The fusion process produces a fine white powder in which any metal ion impurities are combined as the metal silicates. The material has a large surface area of about $1-3 \text{ m}^2\text{g}^{-1}$ and a high porosity, but has only weakly adsorptive properties. It is usually employed with a particle size of 80-100 mesh (*see* **Mesh sizes**) and can take loadings of up to 25% of its own weight of stationary phase.

Cell efficiency. The number of cell volumes required to flush the internal detector cell volume completely. The term is used particularly in connection with HPLC detectors. Poor cell efficiencies (a high number of cell volumes) can produce poor resolution of solutes. Typical values for cell efficiencies are from 2.5 to 9 cell volumes. The better **ultraviolet absorption detectors** have values of between 2.5 and 4 cell volumes.

Cellulose. A white solid polysaccharide obtained either by digesting wood chips or in an almost pure state from cotton. It is still one of the most commonly used materials for chromatography, and in partition systems is employed with either aqueous or organic stationary phases. Acetylated cellulose is employed in **reversed phase systems** and other chemically modified forms are suitable for ion exchange chromatography.

Chabazite. One of the naturally occurring **zeolites**. Chabazite was used in the earliest work carried out on the adsorption of water and alcohols.[63] It has been credited with the formula $Ca_2 [Al_4 Si_8 O_{24}] \cdot 13H_2O$ for the unit cell.

Chelating resins. Resins which involve a matrix formed from polystyrene or similar polymeric materials and are thus similar to the rather more conventional **cation** and **anion exchangers**. To act as a chelating medium the polymeric lattice is substituted with weakly acidic iminodiacetic acid groups,[64] or built upon an ethylenediamine tetraacetic acid structure. The following reaction is characteristic of those used for the more simple type of chelating resin:

Resins of this type are very selective, having a high preference for copper, iron and heavy metal ions and generally acting in a similar manner to ethylenediamine tetraacetic acid (EDTA). They are useful for concentrating trace quantities of metals and for removing trace impurities from other materials.

The range of chelating resins has been greatly extended[65,66] by the incorporation of special functional groups, such as oximes, mercapto and nitroresorcinol, into polymeric structures based upon styrene and/ or divinylbenzene.

C, H, N analysers. Modern instruments for the automatic analysis of organic materials are highly specialized gas chromatographs[67] designed to oxidize the organic compound to carbon dioxide, water and nitrogen. The chemical process is normally carried out in two stages. The initial reaction is oxidation of the sample in a small aluminium boat containing copper(II) oxide catalyst, at 800-1000 °C. Any nitrogen oxides formed are then reduced to nitrogen as the gases are carried by helium into a second reduction furance at 500 °C. Separation of the gases is carried out on a 15 m column with a 3 mm internal diameter using 20% silicone oil as the stationary phase. The percentage composition of the elements is determined from the relative sizes of the peaks obtained, measured by means of a katharometer.

A similar instrument suitable for C, N and S elemental analysis in volatile liquids has also been developed.[68]

Chromathermography (Temperature gradient chromatography). A process in which a temperature gradient is maintained down a GC column. It is an alternative to **temperature programmes** and was first introduced in 1951 to reduce tailing and spreading of bands.[69,70] The process is carried out by slowly moving a heat source (oven) along the column from inlet to outlet. To obtain this type of gradient, Nerheim[71] used as 25 cm glass sleeve wound with a heating tape, producing a gradient of $2\,°C\,cm^{-1}$ down a column. Improved separation efficiency was obtained in the elution of paraffins by passing the sleeve along the column three times at temperatures of 83, 151 and 217 °C.

The procedure does not appear to possess any advantages over temperature programming, but is an obvious alternative where normal programming facilities are not available.

Chromatogram. In GC and HPLC the chromatogram is the resulting output recorded in the form of a continuous chart or computer record obtained from amplifying the detector signal and passing it to a chart recorder and/or computer. This meaning also applies to any form of column chromatography by which the eluate is passed through a detector.

In PC and TLC the chromatogram is the developed surface upon which the separate zones or spots have been revealed, either by spraying with a reagent or by exposure to ultraviolet light.

Chromatograph. Technically, any arrangement consisting of a chromatographic column upon which separations can be carried out. However, the term is usually employed with reference to more sophisticated instruments for GC and HPLC in which the column can be thermostatically controlled or **temperature programmed** and where there are a detector and a record to automatically measure and record the result of the chromatographic process.

The word chromatograph is also used as a verb.

Chromatographic profile. A chromatogram which under specified column conditions shows good resolution of an extensive number of individually related compounds and can be used as a reference for identification purposes. Such profiles are used in conjunction with studies on homologous series such as hydrocarbons, carboxylic acids and closely related cyclic compounds such as steroids.[72]

Chromatographic spectrum. A diagrammatic presentation of R_F values obtained for any single compound by either TLC or PC employing a number of solvent systems. If the R_F values are plotted and joined up in the manner shown in Figure 4, the shape of the polygon obtained is characteristic for that compound using that arrangement of solvent systems.[73] (A chromatographic spectrum has thus nothing to do with electromagnetic radiation — it would be less misleading if it was just called a chromatographic polygon.) If the same process is carried out with a number of compounds it is possible to use the 'chromatographic polygon' as an aid to identification in a similar manner to the finger-printing procedure in infrared spectroscopy.

Chromatographic tank (Development tank). A tank in which paper and thin layer chromatograms are produced. Every form of container from old porcelain drainage pipes to tin cans has been used. However, most modern tanks are made from glass and are either rectangular, large enough to take as 20 cm $\times$ 20 cm TLC plate, or tall and cylindrical (*see* Figure 5, p. 28) to take a quarter plate (20 cm $\times$ 5 cm). For paper chromatography special supports are fitted to carry solvent troughs and support rods suitable for **descending chromatograms**.

Chromatography. Any separative process in which a mixture carried in a moving phase (either liquid or gas) is separated as a result of differential distribution of the solutes between the mobile phase and a stationary liquid or solid phase around or over which the mobile phase is passing. The systems to which this definition applies include all chromatographic processes from paper chromatography to ion exchange and gel chromatography.[74]

The word chromatography was introduced by Tswett[6,7] to describe

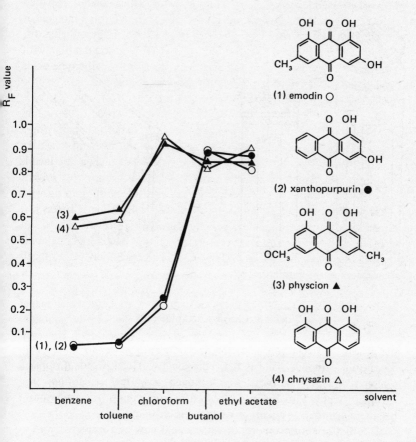

Figure 4. Chromatographic 'spectra' of some hydroxylated anthraquinones

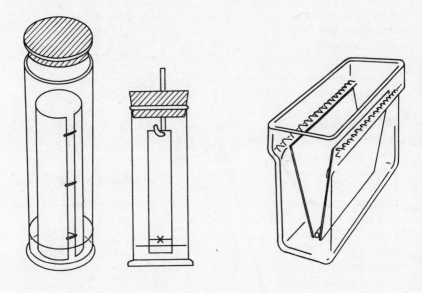

Figure 5. Chromatographic tanks

the process of separation he carried out on plant pigments of a column of calcium carbonate, and is derived from two Greek words *Khroma* (colour) and *grafein* (written). Although very few separations in chromatography actually involve coloured compounds, the name has been retained for all systems related to this technique.

The first scientific reports of what would now be considered chromatography were actually of separations carried out on paper by Runge[75],[76] in 1850. Until 1942 only liquid mobile phases had been employed in chromatographic separations, **gas chromatography** being inroduced in that year for the first time.[77] Since then the most recent advance has been in the development of **high performance liquid chromatography** (HPLC), with many of its features based upon the experience of gas chromatography. *See also* under the name of the individual chromatographic process.

Chromato-sticks. Compressed sticks of adsorbent by which very rapid separations of small quantities of material can be achieved by **adsorption chromatography**. The sample is spotted on to one end of the stick which is placed vertically with the sample end in a shallow depth of solvent. The action is similar to that of TLC, with the solvent ascending the stick and carrying the solutes with it. This type of chromatogram can be readily carried out by using sticks of white chalk.

Chromogenic reagents. Reagents which are sprayed on paper chromatograms, TLC plates and extruded columns to reveal the positions of zones and spots. The reagent will produce a colour either with the individual compounds on the chromatograms or with the surrounding substrate. These chromogenic reagents are sprayed using either a hand atomizer or a can of pressured propellant. The reagent solution must be in the form of a fine aerosol in order to cover the surface uniformly.[78] Typical of the chromogenic reagents employed are: ninhydrin for amino acids, ammoniacal bromothymol blue for lipids, diazotized sulphanilic acid for phenols, and 15% phosphoric acid for steroids. Two general-purpose chromogenic reagents are sulphuric acid and iodine in chloroform.

Chromosorb. The trade name for two series of packing materials for GC.
The Chromosorb 'century' series, listed in Table 3, consists of a range of synthetic polymers that have been used extensively as supports for stationary phases in GLC. In some instances they are also suitable for use as adsorbents in GSC.

Table 3
Chromosorb 'Century' Series

No.	Polymer	Surface area $(m^2 g^{-1})$	Average pore diameter (μm)	Colour
101	styrene/divinylbenzene	< 50	0.3–0.4	white
102	styrene/divinylbenzene	300–400	0.009	white
103	crosslinked polystyrene	15–25	0.3–0.4	white
104	acrylonitrile/divinylbenzene	100–200	0.06–0.08	white
105	polyaromatic	600–700	0.04–0.06	white
106	crosslinked polystyrene	700–800	5 nm	white
107	crosslinked acrylic ester	400–500	9 nm	white
108	crosslinked acrylic ester	100–200	0.03	white

The second Chromosorb series is designated by letters: Chromosorb G is a hard white **diatomaceous earth** support, with a small surface area of 0.5 m² g⁻¹ and pH 8.5 suitable for stationary phase loadings of up to 25%. Chromosorb P is a pink diatomaceous earth like **kieselguhr** and used in a similar manner. Chromosorb A is similar to Chromosorb P but is capable of carrying a heavier stationary phase loading. Chromosorb W is a white diatomaceous earth identical to **celite**, possessing a soft relatively nonadsorptive surface with an area of 1–3 m³ g⁻¹ and pH 8–10. This group of materials is also available in acid-washed and silanized forms. Chromosorb T is a screened teflon with an inert surface useful for the separation of highly polar compounds.

Circular chromatography (Ring chromatography). Both paper chromatography and TLC lend themselves to the production of circular chromatograms by the placing of a spot of mixture at the centre of a sheet or square plate and the addition of solvent to the centre. Capillary action causes a flow away from the centre, the individual components being separated into a series of concentric circles.

Early TLC separations were carried out in this manner[79,80] and a complete apparatus for a wick-fed circular thin layer chromatogram has been devised.[81]

The procedure has been developed into a semiquantitative paper chromatographic method[82] suitable for the simultaneous separation of five different mixtures. This procedure uses a 26 cm paper circle divided into five equal sectors by radial slots cut from near the centre. The five individual samples are each applied at the apex of an arc and solvent is fed by capillary action from a vertical glass capillary support on a pvc dome submerged in the solvent.

Circular TLC is not greatly affected by sample overloading and resolution is improved compared with linear chromatograms. A simple circular paper chromatogram can be run by using two petri dishes (Figure 6), although very sophisticated equipment is now available for this purpose.

Circular columns. In the **Moving bed process** columns for continuous gas chromatography involve the use of quite large quantities of support

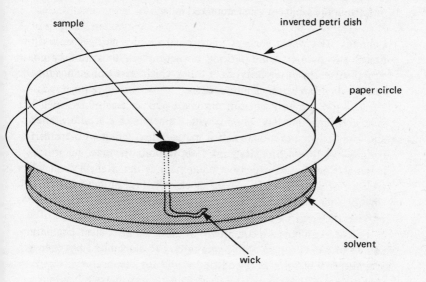

Figure 6. Simple apparatus for circular chromatography

and mobile phase; circular columns have been devised in an effort to obviate this problem. Such columns are shaped into a 1.5 m diameter circle which is continuously turned past a fixed inlet port and two or more fixed outlet ports.[83,84] The less strongly sorbed components travel with the mobile gas phase to one take-off port while the more strongly sorbed components pass with the stationary phase bed in the opposite direction to a second port where they are desorbed by a stripping coil.

Typical conditions employ a powdered firebrick packing and hot nitrogen carrier gas to separate diethyl ether and dichloroethylene at a rate of 2 cm^3 min^{-1} with 97% purity.

Closed strip electrophoresis. *See* **Enclosed strip electrophoresis**.

Co-ions. In ion exchange chromatography, the mobile ionic species possessing the same charge as the **fixed ions**.

Columns. The whole concept of chromatography revolves around the design and nature of the **packing, adsorbent** and/or **stationary phase** contained in the different types of tubes which serve as columns for the variety of chromatographic procedures.

Traditional gravity column chromatography has been carried out almost entirely in glass columns with diameters of several centimetres and lengths of more than half a metre. These columns are still frequently used for chromatographic separations of large quantities of material when more expensive equipment is not available and when time is not particularly important. Such columns can only be used with slightly increased pressures, and operation at elevated temperatures requires the construction of special glass jackets.

For GC, columns are commonly of glass, stainless steel, aluminium, copper or nickel tubing of 2–6 mm internal diameter and 1–4 m length arranged in the form of a coil or U tube for compactness. **Capillary columns** are 100–500 μm in internal diameter and up to 300 m long.

Although glass columns have been used for HPLC at relatively low pressures the columns most commonly employed are stainless steel of internal diameter 1.5–3.0 mm and length 0.5–3.0 m. *See also* **PLOT columns**; **SCOT columns**; **WCOT columns**.

Column chromatography. Any form of chromatography in which a column or tube is used to hold a solid adsorbent, support, or a liquid stationary phase deposited on the column. The term was originally used for liquid–solid adsorption chromatograms and liquid–liquid partition chromatograms run under normal conditions of temperature and pressure in glass tubes. *See also* **Adsorption chromatography** and **Partition chromatography**.

Column performance. The number of **theoretical plates** in the chromatographic column, calculated for each substance from the peak obtained on the chromatogram according to the expression

$$n = 16 \left[\frac{V_R}{w_b} \right]^2$$

Column pressure gradient correction factor. *See* **Pressure gradient**.

Column volume (Bed volume) (X). The total volume of that length of the column occupied by the column packing.

Concentration distribution ratio (D_c or K_c). The ratio of a solute or component when distributed between unit volume of stationary phase and unit volume of mobile phase. Normally this is represented as

$$D_c = \frac{\text{amount of solute cm}^{-3} \text{ of stationary phase}}{\text{amount of solute cm}^{-3} \text{ of mobile phase}}$$

Conductimetric detector. *See* **Electrolytic conductivity detector**.

Conductivity detector. A **bulk property detector** devised for use in HPLC. It consists[85] of a 1.5 μL cell with platinum or stainless steel electrodes forming part of an a.c. or d.c. bridge circuit. Changes of conductivity of the solvent are determined as the solutes are eluted from the column. It is particularly suitable for studying aqueous solutions and can detect 1 part in 10^8 (10 p.p.b.) of electrolytes. The detector is subject to the usual bulk property detector problems of flow and temperature and cannot easily be used with gradient elution. Good temperature control of the cell is necessary as a change of 1 °C gives approximately a 2% change in conductivity. *See also* **Electrolytic conductivity detector**.

Constant composition elution. Elution carried out using a single solvent or a mixed solvent of fixed composition. It suffers from the disadvantage that slow-moving components are often only eluted after extended periods of time and with broad bandwidths. These problems are overcome by **flow programming**, in which the solvent flow rate is progressively increased by **gradient elution**, involving an increase in solvent polarity, or by **temperature programming**, in which the column temperature is steadily increased.

Continuous chromatography. To obtain the rapid separation of large quantities of materials, various techniques have been developed to produce continuous forms of chromatography. The easiest of these is the **batch progress** of automatic sampling and fraction collecting applied regularly to GC and HPLC. Other more truly continuous forms of chromatography have been developed for GC, including the use of **circular columns, moving bed processes** and **radial flow columns**.

Continuous electrophoresis (Preparative electrophoresis). A technique in which a sample is subjected simultaneously to a gravitational flow of buffer and an applied potential at right angles to the buffer flow so as to achieve an electrophoretic separation in which the compounds follow separate paths along the vertical sheet (or curtain)[86] (Figure 7).

The process involves a sheet of filter paper arranged in a manner similar to that for descending paper chromatography (*see* **Descending chromatogram**), its upper edge dipping into buffer solution and its two lower corners also immersed in buffer containing electrodes. The sample is fed by a wick at the top of the curtain at a rate of 0.1–0.2 cm^3 h^{-1}. With an applied potential of 500 V the components follow slightly curved paths to the serrated edge at the bottom where they can be collected into small sample tubes.[87]

Controlled pore glass granules. Porous granules of glass with a high silica content in which the pore size is carefully controlled. They are used to obtain rigid insoluble column packings. Granules with pore size in the range 6–400 nm are particularly suited for processes ranging from **molecular sieving** to **permeation chromatography** and because of their rigidity can withstand the pressures in HPLC columns. *See also* **Aerogels**.

Controlled surface porosity supports. *See* **Porous layer beads**.

Corrected retention time (t_R^0). To allow for the fact that there is a **pressure gradient** along the length of the GC columns, calculation of retention parameters must frequently include a correction (j). Employing this factor, the corrected retention time is obtained by multiplying the retention time by j:

$$t_R^0 = jt_R$$

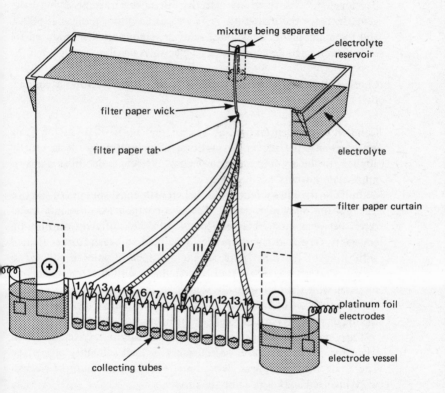

Figure 7. Continuous electrophoresis

Corrected retention volume (V_R^0). A value obtained by the same procedure and for the same reason as that for the **corrected retention time**:

$$V_R^0 = jV_R$$

Coulometric detectors. Quantitative detection for GC can be achieved by coulometry on a nonselective basis if components are pyrolysed to carbon dioxide which is measured in a coulometer.[88] More sophisticated coulometric detectors, including micro designs, have also been developed[89,90,91] which are selective for nitrogen, sulphur and halogens. In these the eluted products are passed into a furnace at 800 °C and the ammonium compounds, acids or sulphur dioxide produced are then measured coulometrically in an electrolytic cell.

Coulson electrolytic conductivity detector. *See* **Electrolytic conductivity detector**.

Countercurrent distribution. A system that is the basis of the **Craig machine** and the parent to **partition chromatography**. It is usually applied to binary mixtures possessing different solubilities for two immiscible solvents.

Initially the binary mixture is shaken with equal volumes of the two solvents and allowed to stand. The two components distribute themselves between the two layers according to their different **distribution constants**. The two layers are then separated and each further treated with more of the other pure solvent and repartitioning is allowed to occur. The process is repeated several times with the result that the material which is preferentially soluble in the lower solvent is concentrated in one set of fractions, and the material preferentially soluble in the upper solvent is in a different set of fractions.

Carrying out this type of separation manually is very tedious, even when the substances differ considerably in their solubility properties. The Craig machine[92] was devised to reduce the amount of manual work involved and to speed up the process.

Counter-ions. In ion exchange procedures, the mobile exchangeable ions in a solution passing down the ion exchange column.

Coupled columns. A technique devised by Snyder[93] to overcome the problems of separating mixtures of components with different polarities by HPLC. By this procedure the sample is initially separated into three parts on a short fore-column, to give less polar, intermediate and most polar fractions. By means of a diverter valve these three fractions are directed to three separate columns possessing different proportions of stationary phase. In some cases the fore-column also serves to remove very strongly sorbed substances. The advantages of the system over gradient elution are claimed to be that, as the same solvent is used throughout, no clearing of a gradient mixture from the system is necessary before reuse of the column. In addition to this any HPLC detector, including those based upon **refractive index**, can be used. In general, however, it is considered to be an inferior technique to **gradient elution** but superior to **temperature programming** and to **flow programming**.

Craig machine. Although the process of **countercurrent distribution** can be carried out by hand it becomes extremely tedious, and to simplify this separative process Craig and Post[94] devised what has become known as the Craig machine. This consists of a series of interconnected tubes, of the type shown in Figure 8, each acting as a separating funnel. Any number of tubes can be employed, from five to several hundred.[95] Initially they are arranged horizontally and the mixture of solutes in one solvent is added to the first tube in sufficient volume not to flow into the next tube when arranged vertically. The remaining tubes each have some of the same solvent placed in them, and then the second partitioning solvent is added to the solution already in the first tube. Partitioning is then brought about by rocking the machine while maintaining the tubes in a roughly horizontal position. On being turned to a nearly vertical position the second (upper layer) solvent containing some of the partitioned solutes flows automatically into the second tube. More of the second solvent is added to the first tube and the process repeated. As a result the lower layers of the first solvent remain in the tubes in which they were originally placed and the upper layers of the second solvent move along the tubes progressively after each separation. The separation of the solute mixtures will depend upon their different **distribution constants**. If the two solvents are in equal volumes in each of the tubes in the Craig machine, the tube number containing the maximum concentration of a particular solute is given by:

$$N_{\max} = \frac{nK_D}{K_D + 1}$$

where n is the number of operations that have been carried out and K_D is the distribution constant. Although partition chromatography is a far more efficient separative process the Craig machine is still employed for the rapid separation of large amounts of binary mixtures in which the two substances possess significantly different distribution constants.

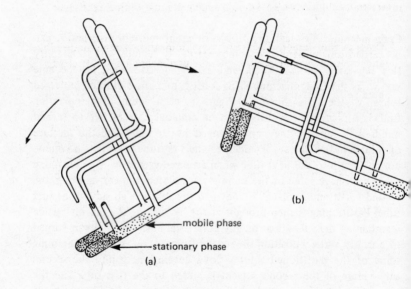

Figure 8. Two interlocking glass units for Craig counter-current distribution. (a) Position during extraction. (b) Position during transfer. (Note: by returning the apparatus from (b) to (a) the transfer is completed. The mobile phase moves on to the next unit and is replaced by a fresh portion.)

Cross-linkage. To form a true three-dimensional matrix in **ion exchange resins** it is necessary to create cross-linkages between adjacent polymer chains. In resins based upon phenol the degree of cross-linkage is determined by the amount of formaldehyde employed to make the polymer, whereas for polystyrene systems the degree of cross-linkage is determined by the proportion of divinylbenzene used in the copolymer: up to 10% divinylbenzene may be employed. Swelling characteristics of the different ion exchange resins are determined by the degree of cross-linkage, and as the cross-linkage is increased so the permeability of the resin decreases. It is, therefore, necessary to establish a balance between the competing requirements of the stability of the resin and its permeability. *See also* **Anion exchanger** and **Cation exchanger**.

Cross-section detector. The first type of **ionization detector** developed[96] and the only one that responds to the permanent gases. It consists of a small ionization chamber in which the carrier gas and solute are irradiated with electrons from a β-ray emitter (^{90}Sr) while a potential is applied across the cell (Figure 9). In the absence of any solute, **baseline** conditions prevail, but when solutes enter the chamber there is a current change arising from the change in total cross-section for ionization. Hydrogen is commonly employed as the carrier gas for this purpose at about 50 cm^3 min^{-1}, but nitrogen and helium may also be used. The current change in the detector is given by

$$\Delta I = \frac{cpVx(Q_{\mathrm{x}} - Q_{\mathrm{c}})}{RT}$$

where p, R and T are the normal gas constants, V is the detector chamber volume, x is the mole fraction of component, and Q_{x} and Q_{c} are molecular ionization constant cross-sections for the solute and carrier gas respectively.

A sensitivity of 10^{-4}–10^{-6} mol and a minimum weight detectability of 0.1 mg have been claimed for the detector. It has been suggested that

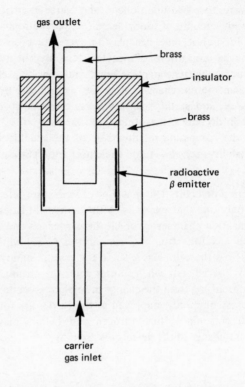

gas outlet

brass

insulator

brass

radioactive
β emitter

carrier
gas inlet

Figure 9. Cross-section ionization detector

this detector should be used as an absolute reference detector for quantitative GC.[97] A micro form employing an 8 μL cell volume has been constructed[98] and is suitable for use with capillary columns.

Cryogenic operation. *See* **Subambient operation.**

Curtain. In **continuous electrophoresis,** the vertical sheet of filter paper employed for the separation. It is cut to a shape that gives a serrated lower edge, acting as tongues at which the separated components can be collected, and side flaps to dip into the buffer solution.[99] *See also* Figure 7, page 35.

Cut and weigh. A most time consuming and only moderately accurate procedure for quantitative measurement in GC and HPLC. It involves cutting out the individual peaks from the chart with scissors and weighing the triangular pieces of paper obtained on an analytical balance. Errors arise from two main causes: incorrect cutting and lack of uniformity in the weight of the paper. Its main advantage as a method of integration is in its economy, but it is not employed very frequently now that **electronic integrators** are so common. *See also* **Triangulation**.

D

Dampening devices (Pulse dampers). Devices that reduce the **noise** arising from the action of reciprocating pumps in HPLC. They consist of capillary tubes, about 6 m long and 0.5 mm internal diameter, in the form of a loosely wound unsupported coil fitted either in line between the pump and the column, or as a blind end at a T junction joined just before the column. The pulses of the pump are absorbed by the free movement of the coil. The presence of dampening devices leads to a larger dead volume in the chromatographic system. *See also* **Pump (reciprocating)**.

Deactivation. Materials employed as column packings, either as **supports** or as **adsorbents**, frequently have an activity greater than that required. Removal or reduction of the activity is achieved by use of one of four general procedures that have been devised for deactivation, which are: (*a*) saturation of the adsorption sites with other active materials that can form hydrogen bonds; (*b*) carrying out a substitution reaction on the active site, for example by **silylation**; (*c*) repetitive washing with acid; (*d*) coating the support with a less active solid material, for example teflon or other polymeric materials.[100]

Dead volume (V_d). An expression rather loosely used over many years but now taken to be the volume between the effective injection point and the effective detection point after deducting the column volume. Thus any section of the chromatographic flow system between the inlet or injection port and the detector outlet in which solute and mobile phase are mixed but not passing over the stationary phase constitutes dead volume in which diffusion of solutes can occur in the absence of separation. In the injection port the dead volume leads to diffusion of the injected sample in the mobile phase before any separation has occurred, hence destroying the characteristics of the **slug injection**. Within the detector the dead volume leads to diffusion of the separated components and a decrease in resolution such that closely spaced components may become merged.

42

Deflection refractometer detector. *See* **Refractive index detector**.

Degassing. Due to the drop in pressure that occurs in HPLC as the solvent passes from the column to the detector, any dissolved gases (most commonly air) are partially released as bubbles within the detector, producing anomalous and misleading results. In general the more polar the solvent the greater its ability to dissolve oxygen. To avoid this problem solvents are degassed before use, either by a nitrogen purge or by heating and stirring under vacuum. Another way of overcoming the problem is by use of a **back-pressure device**.

Deionized water. Water that has been purified by passage through **ion exchange resins**. It is obtained by passing the water first through a **cation exchanger** and then through an **anion exchanger**, or alternatively through a **mixed-bed column**. The chief impurities to be removed from most waters are calcium and magnesium as their carbonates, hydrogen carbonates and sulphates. The only impurities remaining after deionization are small amounts of dissolved gases and any organic impurities washed from the resins themselves.[101] Deionized water has a pH of almost exactly 7, and on evaporation should leave total solids of > 1 p.p.m.; its specific resistance[102] is greater than 1×10^7 ohm cm^{-1}.

The main advantage of the individual column is in the ease of **regeneration**; in the case of the mixed-bed resin the two resins must be separated by reversed water flow before they can be regenerated.

Densitometer. An instrument that greatly aids the making of quantitative measurements in PC and TLC by scanning the individual spots by either reflectance or absorption of a light beam.[103] In the former case the chromatogram is scanned by the moving beam of light (preferably monochromatic) and the intensity of the light reflected from the surface is measured (Figure 10). The difference in the intensity of the light between the adsorbent and any spots are observed as a series of peaks on a trace plotted by a chart recorder. The areas of the peaks correspond to the quantities of the materials in the spots. The alternative procedure produces a similar chart record by measurement of the

Figure 10 here

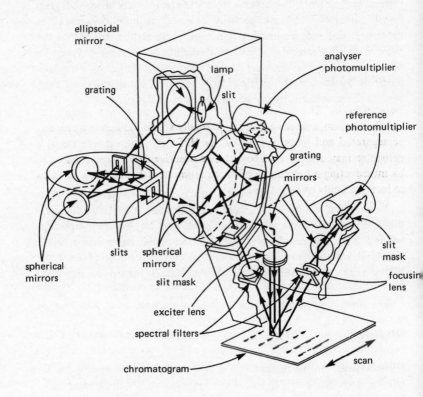

Figure 10. Densitometer for TLC chromatograms (Farrand two-
monochromator visual UV chromatogram analyser)

intensity of the light transmitted through the plate.

In the flying-spot densitometer the individual spots are scanned by a very narrow moving beam and measurements recorded as peaks on chart paper with the sizes corresponding to the concentration of material in each spot. Similarly, with photodensitometers the transmitted or reflected radiation is used to produce a photograph of the chromatogram revealing dark and light zones for the areas of the separated components.[104] The standard deviation for quantitative determinations by densitometry is about 5%.[105]

Derivatization. The process of creating a derivative of a compound, usually to form a more volatile or thermally stable substance which can be separated and identified by GC or HPLC.[106] The preparation of a derivative may also be used to enhance selective detector response, e.g. by introducing a halogen or nitrogen atom into the molecule in order to increase limits of detection.

While the most common form by far of derivatization is that of **silylation** the general term also includes acylation, esterification, formation of oximes etc. Derivatization has become of increasing importance in HPLC,[107] where the range of detectors is more limited than in GC, and reagents such as phenacyl bromide and p-bromophenacylbromide enable ultraviolet active compounds to be formed from a wide range of substances.[108]

Desalting (Group separation). The removal of low molecular weight compounds (both organic compounds and inorganic salts) from those with very much higher molecular weights. It was one of the earliest applications of **gel chromatography**. Success of the method depends upon selecting a gel which will totally exclude the macromolecule required (that is, $K_{av} = 0$) but will permit easy penetration by the small molecule (the salt). Because of the large molecular weight difference, very large sample volumes of the mixture can be 'desalted' before any of the low molecular weight material is eluted from the column. Loadings up to 40% of the total column volume can frequently be made. The procedure has been used for separating sodium chloride from haemoglobin[109] and for purifying nucleic acids.[110,111]

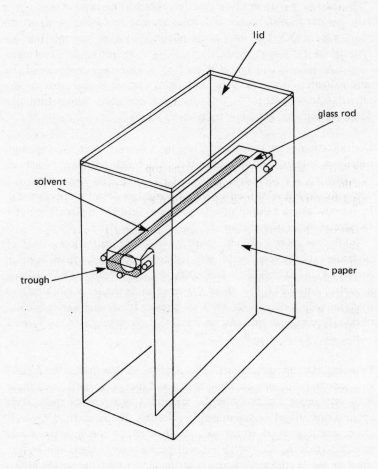

Figure 11. Descending chromatography

Descending chromatogram. One of the two main procedures for running paper chromatograms, and although the most difficult to carry out it gives the better separations. The upper edge of the paper is held in position in a trough of solvent by a rod (Figure 11), and the sample spots which have been placed on a line a short distance from the edge are separated as the solvent flows from the trough due to the combined action of capillary attraction and gravity. Development by descending chromatography is faster than that by the ascending method (*see* **Ascending chromatogram**) and has the added advantage that it is easier to use long strips of paper.

Descending techniques can also be applied to TLC by the use of a wick from the solvent trough to the top edge of the plate, but the method is not frequently employed because of the difficulty of achieving a regular flow from the wick to the plate.

Detectors. Essential pieces of equipment in all GC and HPLC systems that must satisfy a variety of criteria: above all they must be versatile,[112] operate over a wide range of conditions and frequently respond to an extensive range of materials. They may be classified as **differential detectors**, producing a chart record consisting of a series of peaks, or as **integral detectors**, from which the record is a series of steps. Alternatively, they may be classified as (*a*) universal − responding to all or most compounds; (*b*) selective − responding to a group of compounds or a number of species; (*c*) specific − responding to a very limited class of compounds. In all cases a linear response to concentration is desirable. There are two main principles to detection:[113] (*a*) in order to be detected a chemical species must produce a signal; (*b*) the intensity of the signal must be proportional to the quantity of the species producing it.

Although ideally three parameters should be specified to define the response characteristics of all detectors, many values and results in the literature are misleading and ambiguous, so that in many cases the **lower limit of detectability**, the **linearity**, and the **sensitivity** are not known. In addition some detectors are of little value as their response times are too large.

The sensitivity of detectors, σ_d, is related to the lower limit of detectability, E_u, by the equation

$$\sigma_d = \frac{2R_u}{E_u}$$

That is, in order to be detected, the size of the signal should be at least twice the size of the background noise.[114] For GC, sensitivity is also expressed in terms of the **Dimbat–Porter–Stross number**.

There is no ideal detector for either GC[115] or for HPLC[116] and the analyst has to choose the one most suited for his particular purpose at the time. The following detectors are described in the dictionary:

Adsorption (*see* **Thermal adsorption**)
Alkali metal flame ionization (*see* **Thermionic**)
Argon
Beilstein flame
Belt transport
Brunel mass (*see* **Mass**)
Bulk property
Conductimetric (*see* **Electrolytic conductivity**)
Conductivity
Coulometric
Coulson electrolytic conductivity (*see* **Electrolytic conductivity**)
Cross-section
Deflection refractometer (*see* **Refractive index**)
Dielectric constant
Disc conveyor flame ionization
Discharge spectra
Electrical discharge spectra (*see* **Discharge spectra; Electrodeless discharge**)
Electrochemical
Electrodeless discharge
Electrolytic conductivity
Electron capture
Electron mobility
Flame emission (*see* **Photometric**)
Flame ionization
Flame photometric (*see* **Photometric**)
Flame thermocouple

Fluorimetric
Fresnel refractometer (*see* **Refractive index**)
Gas density balance
Gas-flow impedance
Gas volume (*see* **Janak's method**)
Heat of adsorption (*see* **Thermal adsorption**)
Helium
Hot wire (*see* **Katharometer; Thermal conductivity**)
Hydrogen flame (*see* **Flame thermocouple**)
Infrared
Ionization cross-section (*see* **Cross-section**)
Ionization
James–Martin gas density balance (*see* **Gas density balance**)
Janak's method
Katharometer
Mass
Microcoulometric (*see* **Coulometric**)
Microwave plasma
Moving wire (*see* **Transport**)
Phase transformation (*see* **Wire transport**)
Photoionization
Photometric
Piezoelectric
Polarographic
Potential difference radio-frequency
Radioactivity
Radio-frequency
Refractive index
Selective
Solute property
Sorptiothermal (*see* **Thermal adsorption**)
Surface potential
Thermal adsorption
Thermal conductivity (*see* **Katharometer**)
Thermal (*see* **Thermal adsorption**)
Thermionic

Transport
Ultrasonic
Ultraviolet absorption
Wire transport

Development. Traditionally, the process of carrying out a chromatographic separation. Use of the word in this sense is, however, being discouraged because of the confusion with photographic development. It is now more usual to talk in terms of running a chromatogram, and the various procedures for this are described under **displacement chromatography**, **elution chromatography** and **frontal analysis**.

Development tank. *See* **Chromatographic tank**.

Dextran gels. Dextran is a water-soluble macromolecule consisting of chains of glucose joined together mainly by α-1,6-glycoside linkages. The insoluble dextran gels[117] (*see* **Sephadex**) are formed by cross-linking dextran chains with glycerol ether groups obtained by reacting the dextran with epichlorohydrin.[118]

Diagonal chromatography. The technique whereby, if a paper or thin layer chromatogram is run twice with the same solvent system in directions at right angles to produce a two-dimensional chromatogram, the various components will be separated out along a diagonal. This process is used if it is believed that one of the components is decomposing or reacting during the chromatographic process. If no secondary product is being formed then the diagonal distribution is obtained, but where one of the compounds is undergoing chemical reaction with either the solvent or the adsorbent then a spurious spot not falling on the diagonal line is also obtained.

Dialysis. The use of animal membranes, cellulose films, agar membranes and ion exchange sheets as semipermeable membranes permitting passage only of selected molecules or ions in aqueous solution.[119,120] The action of the membranes is such that, whereas passage of water or small molecules through them is possible, the larger electrolytes in solution are prevented from passing through. This results in a concentration gradient between the solution on one side of the membrane and solvent on the other. There is actually competition between the kinetic movement of the solute molecules or ions on the one hand and the osmotic pressure differences on the other, with the membrane serving to prevent the former movement from occurring.

With ion exchange membranes the main application is in the desalination of brackish water by electrodialysis.[121] The equipment employed for electrodialysis consists of a number of chambers separated by ion exchange membranes with brackish water passing between them. By the application of a potential across the whole length of the chambers, sodium chloride becomes concentrated in each alternate cell while the water is correspondingly less brackish in the other alternate cells. By using several hundreds of cells it is possible to produce large volumes of drinking water from water containing up to 4000 p.p.m. of sodium chloride.

Diatomaceous earths. The remains of microscopic single-celled organisms (diatoms) in the form of amorphous hydrated silica containing a small proportion of metal oxide impurities. **Celite** and **kieselguhr**, the most common **supports** used in partition chromatography, are both diatomaceous earths.[122] The white diatomites are obtained by heating the excavated diatoms with a little sodium carbonate to above 900 °C. Pink diatomites are also obtained by heating to 900 °C, but with the exclusion of any other material, and the resulting pink colour is believed to be due to the presence of iron oxide.

The diatomaceous earths possess surface areas of approximately 20 m^2 g^{-1} and are very porous. They are not completely inactive and some interaction with solutes occurs. The main groups on the support surfaces are Si—OH and Si—O—Si. *See also* **Chromosorb**.

Dielectric constant detector. A detector which is based upon the measurement of capacitance of the eluant between parallel plates in the detector cell. It has been applied to both GLC[123,124] and HPLC.[125] This type of detector is only sensitive to polar compounds, and to obtain good results the difference in dielectric constant between the solute and mobile phase should be as large as possible. Although it has not been used generally, the detector has been found to be useful for permanent gas analysis.[126]

Differential detector. According to Huber,[127] a detector which responds to an input signal which is proportional to the differential quotient of the sample amount with respect to time or volume. Mathematically this is expressed as

$$X = a_i \frac{dQ}{dt}$$

where a_i is a proportionality factor for component i. The resulting output from this type of detector is a signal which produces a series of peaks on a chart, the areas of which correspond to the quantity of material which has passed through the detector (Figure 12). Most detectors used in chromatography are of the differential type but can be linked to **electronic** and **disc integrators** to produce a direct measure of the peak areas and the quantity of solutes. *See also* **Integral detector**.

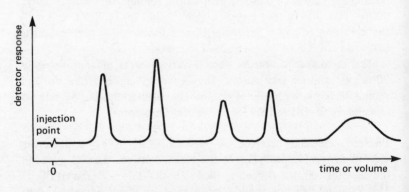

Figure 12. Signal produced by a differential detector

Dimbat–Porter–Stross number. The value of detector **sensitivity** for GC expressed[128] in terms of cm^3 mV^{-1} mg^{-1} by use of the equation

$$S = \frac{A_i C_1 C_2 C_3}{w_i}$$

where A_i is the peak area in cm^2, C_1 is recorder sensitivity in mV cm^{-1}, C_2 is the reciprocal chart speed in min cm^{-1}, C_3 is the flow rate at the column exit, and w_i is the weight of sample in mg.

The Dimbat–Porter–Stross number increases in value with increased sensitivity of the detector. For a **katharometer** the value is about 14 000 and for a **flame ionization detector** it is about 20 000.

It should be noted that this is a simplified form of the sensitivity equation and relates sample weight to peak area.

Dimethyldichlorosilane. A reagent, usually referred to as DMDCS, which is used for **deactivation** to reduce the number of adsorption sites on siliceous supports in GC.[129] The reagent converts the hydroxyl groups on the support material, which cause tailing of peaks, into inactive silicates[130] by the following route:

Any unreacted chlorine atoms on the silyl groups are removed by treating the deactivated support with methanol.

The reagent has also been employed for **silylation** to produce volatile derivatives, suitable for GC separation, from involatile organic compounds. *See also* **Hexamethyldisilazane.**

Dinonyl phthalate. A diester of phthalic acid, usually referred to as DNP. The diesters of phthalic acid were among the earliest materials employed as stationary phases in GLC and DNP is probably the most frequently used of these compounds. Being a diester, with the structure

$$\text{CO}_2\text{C}_9\text{H}_{19}$$
$$\text{CO}_2\text{C}_9\text{H}_{19}$$

it is semipolar in character and suitable for separations of alcohols, aldehydes, esters and ketones. It is used up to a 10% loading on the support and has a maximum operating temperature of 150 °C.

Disc conveyor flame ionization detector. A modified form of **transport detector** in which the transportation is carried out with a porous disc made from machinable alumina.[131] The HPLC eluate is deposited on the disc and the solvent evaporated by a heater/exhaust fan system. The edge of the disc is enveloped by hydrogen flames from dual nozzles on either side of the disc, and ions formed in the **plasma** are collected at the nozzles and at a semicircular collector electrode. The detector gives a linear response and has a minimum detection limit of approximately 2 ng s^{-1}. Earlier forms of this detector have been described employing either a brass disc[132] or a disc-shaped wire net[133] to collect the effluent from which the solvent is then evaporated. *See also* **Belt transport detector** and **Wire transport detector**.

Disc electrophoresis. A variation on conventional gel electrophoresis, first described[134,135] in 1961. For this purpose the gel tube is placed vertically and filled with one or more small-pore-size gels which serve to produce a partial molecular sieving effect. In early systems three layers were employed: (*a*) the sample gel, formed in situ from the protein sample and a gel monomer, the polymerization occurring after it was added to the column head; (*b*) a large-pore stacking gel through which the protein moved rapidly, leading to bunching occurring at the interface with (*c*) the fine-pore separating gel. Stacking and resolution were improved by using a stacking compartment buffer of lower pH than that in the separating gel so that proteins moved rapidly away from the interface once they entered the separating gel, forming well-established discs in the column. In more recent approaches[136] it has been shown that it is possible to dispense with one or both of the first two gels with the separation depending mainly upon the nature of the separating gel.

Discharge spectra detector. Several detectors which involve the use of electrical discharges in one form or another have been designed.[137,138] In this particular type of detector a d.c. discharge through the helium carrier gas is used to produce emission spectra of eluates from GC columns.[139] The discharge causes fragmentation of the solute molecules and the resulting characteristic spectra of C, H, F and of CH and C_2 species are studied. The observed spectrum changes as individual solutes are eluted. The detector is most suitable as a selective detector for halogenated compounds. *See also* **Electrodeless discharge detector**.

Disc integrator. One of the early automatic integration methods for chromatographic peaks that used the ball and disc principle (Figure 13). It is a mechanical device in which a ball placed on a rotating horizontal disc spins at a speed proportional to its distance from the disc centre. The disc is made to rotate at a constant rate and the position of the ball is changed whenever the recorder pen moves away from the baseline to trace a solute peak. The rotating ball is used to drive a roller which by means of a spiral cam moves a second pen to produce an integrated area in the form of a zig-zag line at the bottom of the chart paper. The number of pen movements under a particular chart peak is proportional to the area of the peak. The precision of this integration method is only bettered by **electronic integrators**.

Displacement chromatography. A chromatographic process in which the running of the chromatogram is carried out by placing the mixture to be separated on the column and then eluting with an eluent containing a substance which is more strongly sorbed than are the components of the original mixture. The result is that although the individual components of the mixture are separated according to their own individual

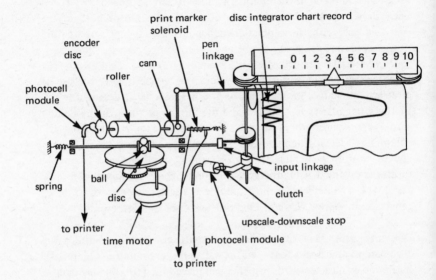

Figure 13. Disc integrator

properties, they are displaced by each other and by the more strongly
sorbed solvent. This leads to a bunching up of the zones, as shown in
Figure 14, so that they are eluted with fairly sharp and distinct boun-
daries between each other.

This type of column operation is used most frequently in **ion exchange chromatography**. *See also* **Elution chromatography** and **Frontal
analysis**.

Distribution coefficients. Several distribution coefficients have been
defined for general application in chromatography. They differ mainly
in the way in which the quantity of individual solute in the stationary
phase is specified. The three most important forms of distribution
coefficient are:

1. for stationary phases in which swelling occurs, such as gel chromatography and ion exchange,

$$D_g = \frac{\text{amount of solute per gram of dry stationary phase}}{\text{amount of solute per cm}^3 \text{ of mobile phase}}$$

2. where the weight of the solid phase cannot be readily obtained then it is measured in terms of the 'volume distribution coefficient' given by

$$D_V = \frac{\text{amount of solute in the stationary phase per cm}^3 \text{ of bed volume}}{\text{amount of solute per cm}^3 \text{ of mobile phase}}$$

3. where surface activities are concerned

$$D_s = \frac{\text{amount of solute per m}^2 \text{ of surface}}{\text{amount of solute per cm}^3 \text{ of mobile phase}}$$

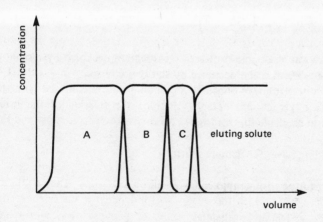

Figure 14. Displacement chromatography

See also **Distribution constant** and **Mass distribution ratio**.

Distribution constant (K_D, Formerly Partition coefficient). The ratio of the amount of solute per unit volume of stationary phase to the amount of solute per unit volume of mobile phase. It is a measure of the distribution of a substance between the two phases in the chromatographic system. Various symbols have been used in the literature for this term, including k' and α, but the recommended symbol is now K_D such that

$$K_D = \frac{\text{mass of solute in unit volume of stationary phase (g cm}^{-1})}{\text{mass of solute in unit volume of mobile phase (g cm}^{-1})}$$

For GLC it can be shown that

$$K_D = \frac{V_N}{V_L} = \beta k$$

and that the distribution constant is related to the R_F **value** by the relationship

$$K_D = \frac{a}{b}\left(\frac{1}{R_F} - 1\right)$$

where a is the fraction of area cross-section occupied by mobile phase and b is the fraction occupied by stationary phase.

The distribution constant K_D should not be confused with the K_d used in gel chromatography defining the fraction of the inner gel volume available for diffusion.

Dividing valve. *See* **Sample splitter**.

DMDCS. *See* **Dimethyldichlorosilane**.

DNP. *See* **Dinonyl phthalate**.

Donnan membrane potential. When a sphere of water-wet acidic **cation exchanger** is immersed in a solution of hydrochloric acid (or other strongly ionized acid), no ion exchange takes place, but the hydrochloric acid diffuses into the water within the resin matrix until an equilibrium is established. The hydrochloric acid concentration inside the resin, however, is always less than that outside the matrix due to the electrical potential across the interface between the ion exchanger and the solution. This is known as the Donnan membrane potential, and it exists between each ion exchange resin particle and the surrounding solution.[140]

Dry-column chromatography. A system of techniques which are a direct translation of TLC separative procedures into a preparative scale on a column.[141,142] The column for this process is dry packed with adsorbent, usually **alumina** or **silica gel**, and the mixture is deposited on top of the column. Solvent is then added to the head of the column and percolates downwards due to the capillary action and the force of gravity. Separation is complete within about thirty minutes. By using columns of diameter about 4 cm and length about 50 cm, loadings of 6–7 g of material for separation can be made. Both glass and nylon tubes have been used for this purpose, the latter having the advantage that after development the bands of solute can be readily cut from the tube without resorting to extrusion.

Dual layer chromatography. A form of two-dimensional TLC in which the separation is carried out in a strip of adsorbent in the first direction and into a layer of a different adsorbent in the second dimension.[143] For this purpose the TLC plate must be coated as shown in Figure 15. The procedure, which is sometimes called multiple layer chromatography, has the advantage that by using two different adsorbents with two different solvent systems it is possible to obtain totally different movements of solutes in the two dimensions.

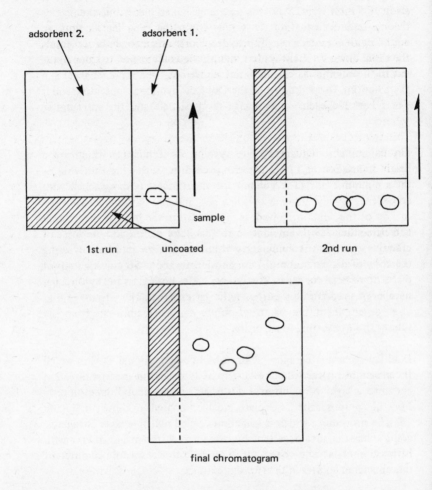

Figure 15. Dual layer chromatography

E

Eddy diffusion term. The first term in the common representation of the **van Deemter equation**[144]. It is the contribution to the value for the **height equivalent to a theoretical plate** made as a result of molecules travelling along different paths through the column due to the particle size of the column packing. Its value is given by

$$A = 2\lambda d_p$$

This term does not depend upon the solute, solvent or operating conditions of the column and is usually the smallest contributor to the height equivalent to a theoretical plate. In the original presentation of the equation[145] the term was called the axial eddy diffusion and placed second.

Effective theoretical plate number (N). The theoretical number of effective plates in a column gives a truer idea of the real efficiency of the column for separative purposes than does the number of theoretical plates, as it corrects for **dead volume**. The value of N is related to the number of theoretical plates, n, by the relationship

$$N = \left(\frac{k}{1+k}\right) n = 16\left(\frac{t_R - t_M}{w_b}\right)^2$$

It can also be related to the resolution, R_S, by the equation

$$N = \frac{16 R^2_S}{(1-\alpha)^2}$$

Efficiency of a column. A measure determined by the number of theoretical plates, n, in the column, calculated from the equation

$$n = 16\left(\frac{d_R}{w_b}\right)^2$$

More exactly the equation should involve the **effective theoretical plate number** and the correct retention distance, in which case it becomes identical with the **separation factor**, S:[146]

$$N = 16\left(\frac{d'_R}{w_b}\right)^2 = S$$

Effluent splitter. A means of achieving division of the effluent from a column to give two or more separate streams (Figure 16). Such a system is needed in cases in which two separate detectors are used simultaneously, or when simultaneous collection and detection of fractions is carried out. In many cases the effluent splitter consists of a needle valve which can divide the stream in various proportions from a 1:1 ratio to 1:100. *See also* **Sample splitter.**

Electrical discharge spectra detector. *See* **Discharge spectra detector** and **Electrodeless discharge detector.**

Electrochemical detector. Any detector based upon electrochemical principles,[147] including amperometry, conductimetry, coulometry and polarography. In one particular form of electrochemical detector designed for LC the eluate is passed through a thin layer cell, of about 1 μL volume, in the presence either of a glassy carbon or carbon paste electrode. Only compounds which are electroactive at selected potentials are detected and as such the detector is mainly selective for organic compounds possessing major polar groups. The reference electrode for the circuit is Ag/AgCl.

Electrode. In any electrolytic cell, either of the two conductors by which the current enters or leaves the cell. The electrode attracting negatively charged particles is the **anode** and that attracting positively charged particles the **cathode**. In apparatus used for electrophoresis these two electrodes are usually identical and interchangeable.

Electrodeless discharge detector. To overcome the problem of contamination of the electrodes which occurs with detectors employing

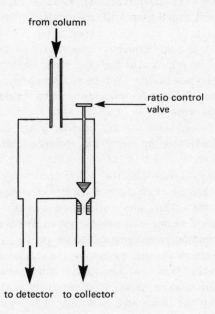

from column

ratio control
valve

to detector   to collector

Figure 16. Effluent splitter

electrical discharges through the effluent gases[148] from gas chromatographs, attempts have been made to isolate the gas stream from the
electrodes.[149,150] The detectors developed achieve electrical conductivity by subjecting the gas to a discharge, but contact between the
vapour and the electrodes is avoided by producing the discharge in the
annulus of coaxial glass tubes. A discharge space of approximately 12
cm^3 is used with an excitation potential of 9.0 kV at 70 °C. It is
claimed that quantities as small as 10^{-7} mol can be detected. The
detector is suitable for studies on permanent gases and many organic
compounds. *See also* **Discharge spectra detector**.

Electrodialysis. *See* **Dialysis**.

Electrolytic conductivity detector (Coulson electrolytic conductivity detector; Conductimetric detector). A detector[151] in which organic material eluted from a GC column is combusted in excess oxygen. The products of combustion are dissolved in water in a d.c. conductivity cell[152] and a continuous measurement of the conductivity is carried out. The response of the detector is linear with concentration and it can be made specific for halogens, nitrogen and sulphur.[153] It is suitable for reliable pesticide analyses but, like the **flame ionization detector**, it is destructive.

The cell used for this purpose is best constructed with platinum electrodes forming part of a d.c. bridge circuit.

Electron capture detector. A detector[154] which was first developed[155] in 1960 for the particular purpose of detecting minute concentrations of pesticide residues, and is often used in parallel with another detector.

The detector uses low-energy electrons from either a tritium or ^{63}Ni source to ionize the carrier gas in an ionization chamber at a potential just strong enough for the collection of all the free electrons produced. When an electron-capturing solute enters the chamber a current decrease occurs. The sensitivity of the detector (Figure 17) is greatest for compounds possessing halogen atoms and oxygen atoms. The carrier gas is nitrogen or argon containing 1% carbon dioxide, enabling a sensitivity of 10^{-12} mol and a minimum weight detectability of 0.05 μg to be achieved with pesticides.[156] Application of the detector for determining trace quantities of polycyclic aromatic hydrocarbons has been achieved[157] by incorporating 2% oxygen in the nitrogen carrier gas.

Electronic integrators. The most expensive integrators for use in chromatography but also the most accurate when used correctly, producing a digital read-out and/or print-out of the integrated areas. They are designed to operate as soon as there is any movement away from the baseline. In some instruments the integrator can also be made to compensate for **baseline drift** and can be adjusted to provide corrected readings for overlapping peaks. *See also* **Disc integrator**.

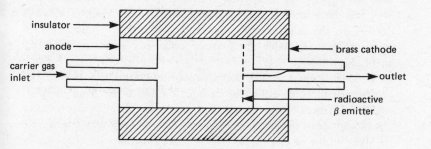

Figure 17. Electron capture detector

Electron mobility detector. A detector which is used for studying permanent gases and is based upon changes in the mobility of electrons[158] arising from the presence of solutes in the carrier gas (usually argon). It is a modified form of the **argon detector**. Any permanent gas, such as oxygen, in the carrier gas causes a drop in electron energy and a reduction in the number of noble gas atoms excited to the metastable state. This causes a drop in the current flow corresponding to the concentration of the permanent gas.[159]

Electro-osmosis. The movement of a liquid relative to a stationary charged surface which results from an applied electric field. In the case of aqueous solutions the movement of the water is towards the cathode, thus assisting movement of positively charged particles and opposing the movement of negatively charged particles. As a result the occurrence of electro-osmosis can have a serious effect on electrophoretic separations. It can, similarly, have an adverse effect on electrodialysis with ion exchange membranes. *See also* **Dialysis** and **Electrophoresis**.

Electrophoresis. One of the four related electrokinetic phenomena involving the relative movement between the two parts of an electric double layer,[160] namely the movement of the charged surface (in this case colloidal or dissolved substances) relative to a stationary **buffer solution** arising from an applied electric field. The term electrophoresis was first introduced in 1909 by Michaelis[161] in connection with his studies on the migration of colloids. The whole technique has since become of considerable importance particularly in biochemical and medical studies.

Depending upon the net charge carried by the particles or ions, the migration will be to either the cathode or the anode. Separation between substances occurs from different rates of migration as a result of the magnitude of the net charge and molecular or ionic dimensions.

Many forms of electrophoresis have been carried out in gel media and this has also been studied under the influence of zero gravity on the Apollo 14 and Apollo 16 space missions.[162]

The term zone electrophoresis is applied to those systems in which ionic mobilities are studied on strips of paper, cellulose acetate or acrylamide. In zone electrophoresis the strip support material is soaked in the buffer solution and the mixture to be separated applied as a spot or band across the centre. A potential is then applied between the two ends of the strip, causing the components to be attracted to one or the other end of the strip according to their charges, sizes and shapes[163] (Figure 18). The separations are most commonly carried out on strips 3–5 cm wide and 40 cm long under either low voltage conditions (<20 V cm^{-1}) or at high voltages (>20 V cm^{-1}), the latter having the disadvantage of producing substantial heating effects. *See also* **Enclosed strip electrophoresis** and **Immersed strip electrophoresis**.

Thin layer electrophoresis is also now a common procedure involving electrophoretic separations on thin layers impregnated with electrolytes. This may also be used for two-dimensional separations in which electrophoresis is applied in one dimension and chromatography in the second dimension.

See also **Electro-osmosis**; **Sedimentation potential**; **Streaming potential**.

Electrophoretic mobility (u_E). Movement of particles and substances in solution under electrophoretic conditions is influenced by a number of factors, and the overall combined effect is expressed in terms of the

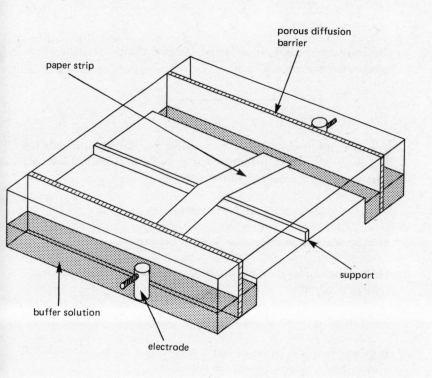

Figure 18. Apparatus for zone electrophoresis

electrophoretic mobility by the equation

$$u_E = \frac{Q_E}{6\pi\eta r}$$

This may be re-expressed in terms of the dielectric constant of the medium and the zeta potential:

$$u_E = \frac{\zeta \epsilon_r E}{6\pi\eta}$$

The electrophoretic mobility is therefore proportional to both the zeta potential and the dielectric constant of the medium, and inversely proportional to the viscosity of the medium.

Elemental analysis. *See* **C, H, N analysers.**

Element selective detectors. *See* **Selective detectors.**

Eluate. The combination of mobile phase and solute following passage through a column:

eluent + solute = eluate

It should, however, be noted that it is becoming the practice in HPLC to refer to the effluent rather than the eluate.

Eluent. The mobile phase (gas, liquid, mixed solvent) used to carry out the chromatographic separation.

Eluotropic series (Mixotropic series). In liquid–liquid partition chromatography the distribution of a solute between the two phases is dependent upon the polarities of the two phases as well as the nature of the solutes. Similarly, the rate of elution from an adsorption column is greatly dependent upon the polarity of the solvent employed. Because of the relationship between speed of elution and solvent polarity, many attempts have been made to arrange solvents in eluotropic series of either increasing or decreasing polarity.[164] Such series are either empirical or based upon theoretical concepts such as the **solubility parameters** introduced by Hildebrand.[165] It should be noted that a series of solvents arranged for one partition or adsorption system is not necessarily completely the same for another.[166]

A typical general purpose series of solvents in order of increasing polarity is: pentane, hexane, diethyl ether, carbon tetrachloride, ethyl acetate, toluene, chloroform, tetrahydrofuran, benzene, acetone, dichloromethane, chlorobenzene, anisole, dioxan, methyl iodide, carbon disulphide, propanol, pyridine, nitrobenzene, ethanol, phenol, dimethylformamide, acetonitrile, acetic acid, dimethyl sulphoxide, methanol, ethanolamine, ethylene glycol, formamide, water.

Elution. The process of passing the chromatographic **mobile phase** over the stationary phase and at the same time transporting the solutes. There are three general methods of carrying out elution: **displacement chromatography**, **elution chromatography** and **frontal analysis**.

Elution chromatography. One of the three methods used for carrying out chromatographic separations, and by far the most common. The mixture is applied as a small quantity at the head of the column or as a spot on paper or thin layer and the individual components are separated by being transported along the stationary phase by the continuous addition and movement of the mobile phase. Separation of the components occurs as a result of their different chemical and/or physical properties. This type of development produces a chromatogram in which the solutes give either separate spots (as in PC and TLC) or a series of peaks on chart paper (GC or HPLC) (Figure 19). *See also* **Displacement chromatography** and **Frontal analysis**.

Elution volume (V_e). A term used almost exclusively in **gel chromatography** with reference to the volume of solvent measured between the placing of a substance at the head of the column and its elution at maximum concentration at the bottom. For a totally excluded substance like **Blue dextran** the elution volume is identical with the

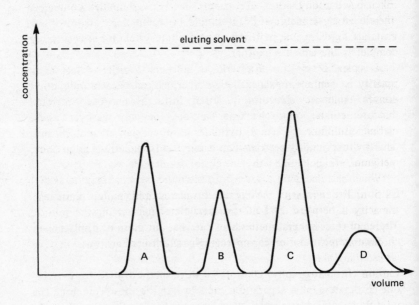

Figure 19. Elution chromatography

interstitial volume, that is, in this particular case $V_e = V_o$.

The elution volume can be considered comparable with the **retention volume** employed in other forms of chromatography.

Enclosed strip electrophoresis. One of the methods used to overcome **capillary flow** in zone **electrophoresis**. The procedure has been widely applied and involves enclosing the strip, except for the two ends dipping into the **buffer solution**, between two cooled metal plates.[167] Various modifications have been made to the basic procedure, including the use of a uniform pressure bag on the plates to prevent zone distortion, and paper wick connections between the electrophoresis strip and the buffer.[168] *See also* **Immersed strip electrophoresis**.

Exchange capacity. Although it is normal to talk in general terms about the exchange capacity, this can be very misleading. The term may, for instance, refer to the total capacity of the **ion exchange resin** – meaning the total number of exchangeable sites, expressed on a dry-weight basis as so many milliequivalents per gram. However, as many of the sites in a resin are not available for exchange, due to the inability of ions to penetrate fully the resin matrix, it is more correct to use the term in connection with the **available capacity** of the resin under defined conditions. These conditions are normally on the swollen resin and are expressed on a wet-weight basis in terms of milliequivalents per gram.

When calculated on a dry-weight basis the capacity may be as high as 5 milliequivalents g^{-1}, while on a wet-weight basis the available capacity is between 1 and 3 milliequivalents g^{-1} depending upon the degree of cross-linking in the resin structure.

Several forms of exchange capacity have therefore been defined, as listed in Table 4.

Table 4
Ion Exchange Resin Capacities

Theoretical specific capacity, Q_o	meqs of ionogenic group per gram of dry ion exchanger
Practical specific capacity, Q_A	meqs or millimoles of ions exchanged per gram of dry ion exchanger
Volume capacity, Q_V	meqs of ionogenic group per cm^3 of swollen ion exchanger
Breakthrough capacity, Q_B	the practical capacity in meqs, millimoles or mg per g of dry ion exchanger or per cm^3 of bed volume obtained experimentally
Bed volume capacity	meqs of ionogenic group per cm^3 of bed volume

Exclusion. Molecules which are not sorbed and are unable to penetrate the interstices of **molecular sieves**, **aerogels** and **xerogels** are said to have been excluded. The exclusion process forms the basis of **gel chromatography** in which separations are achieved as a result of the different abilities of the species to penetrate the gels. Totally excluded substances, such as **Blue dextran**, move with the solvent front and are used to determine the **interstitial volume** of the column.

Exclusion chromatography. One of a variety of names for **gel chromatography** which was suggested[169] in 1962 but has not been generally accepted.

Exclusion limit. A term used in **gel chromatography** to express the molecular-weight limit for separations that are possible on the particular gel. It may be defined as the molecular weight above which molecules are incapable of penetrating the gel pores and are eluted with the **interstitial volume** of the column. Molecules with molecular weights below the exclusion limit are able to penetrate the gel matrix and to be separated according to their molecular dimensions. *See* Figure 43, page 169.

F

FID. *See* **Flame ionization detector.**

Fixed ions. The nonexchangeable ions held in an **ion exchanger** which are opposite in charge to the **counter-ions** in solution.

Flame emission detector. *See* **Photometric detector.**

Flame ionization detector(FID). Probably the most widely used of all GC detectors, combining a high sensitivity with a response to all organic compounds. It was first described[170] in 1958. It is based upon the increased current obtained as a result of burning the column effluent gas with hydrogen and air at a small jet with an applied potential between this and the collector electrode. The burning of solutes in the flame leads to improved conductivity through the ions created and the resulting increased current is used to actuate a chart recorder.

The detector[171] is normally constructed in stainless steel and has an effective volume of $2 \mu L$. The collector electrode is situated 0.5-1.0 cm above the 0.5 mm orifice (Figure 20). It has a sensitivity of 10^{-9} to 10^{-12} mol and a minimum weight detection limit of about 0.1 μg, but does not respond to oxygen, nitrogen, carbon monoxide, carbon dioxide, carbon disulphide, hydrogen sulphide or sulphur dioxide. Because of its high sensitivity it is particularly useful for work in conjunction with capillary columns.

The Beilstein flame detector, photometric detector, flame thermocouple detector and thermionic detector are all variations on the FID, while the various transport detectors for HPLC all require an FID for their operation.

Flame photometric detector. *See* **Photometric detector.**

Flame thermocouple detector. A GC detector in which the column effluent is destroyed by combustion in a hydrogen flame at a small glass capillary. The device[172] measures changes in temperature of the

73

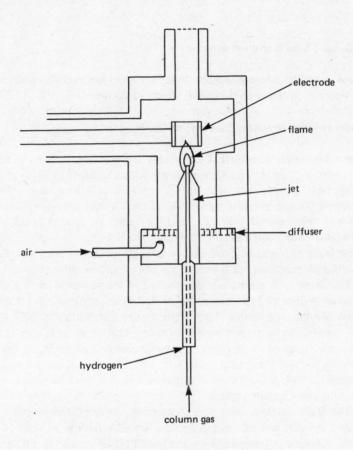

Figure 20. Flame ionization detector

flame due to the variation of composition arising from the presence of solutes in the carrier gas. The temperature change is detected by a thermocouple, made from platinum, palladium, iridium or gold, and located just above the flame, and the response is related to the mass of material combusted in the flame[173]. It requires either hydrogen or a 3:1

hydrogen–nitrogen mixture as the carrier gas. This detector has a very small **dead volume** but is less sensitive than the **flame ionization detector.**

Flash pyrolysis. One of the most widely used procedures for pyrolysis in GC in which the temperature of the sample is raised in a fraction of a second to the pyrolysis temperature required for the material. The temperatures employed are between 800 and 1000 °C, and by using standard conditions the pyrolytic products give a chromatogram which serves as a fingerprint for the original material. *See also* **Pyrolysis gas chromatography.**

Flory-Fox relationship. The molecular-weight distribution for separations on gel columns is influenced by a number of factors involving molecular dimensions and shapes in addition to molecular weights. Various attempts have been made to interrelate the contributions of these factors. One of the more successful approaches has been the use of the hydrodynamic size of the molecules. The Flory-Fox relationship[174] shows that the intrinsic viscosity $[\eta]$ may be regarded as a measure of the ratio of the effective hydrodynamic volume of the polymer to its molecular weight, such that

$$[\eta] \propto \frac{R_e{}^3}{M_r}$$

for long-chain lengths, where R_e is the hydrodynamic radius of the molecule.

The intrinsic viscosity can be further related to the mean-square end-to-end molecular distance $(\bar{r})$ and the molecular weight by

$$[\eta] = \Phi \frac{(\bar{r}^2)}{M_r}$$

where Φ is a constant for all polymers regardless of solvent and temperature.

These concepts are employed[175] in predicting and calculating separations of long-chain compounds in gel systems. *See also* **Mark-Houwink equation.**

Flow cell. Because of the importance of the **ultraviolet absorption detector** in HPLC and in **LC/IR interfaces,** considerable study has been made on the design of the flow cells in order to minimize the noise and diffusion. The classical design is a Z shape,[176] but rectangular cells with internal baffles have also been used[177] (Figure 21). Similar studies have been applied to cells for electrochemical detectors in which a wide range[178] of shapes and sizes is employed.

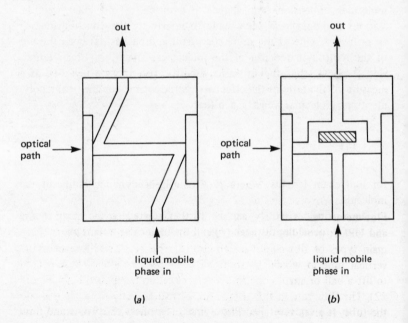

Figure 21. (a) Z flow cell; (b) II flow cell

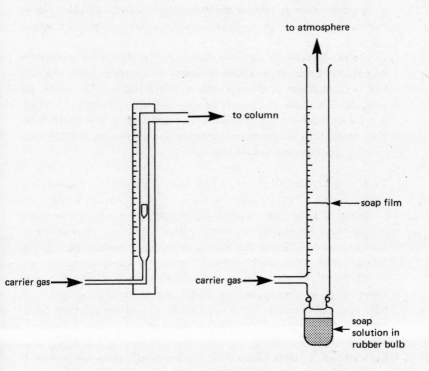

Figure 22. Flowmeters

Flowmeter. An essential part of the necessary equipment used in GC and HPLC, providing reproducible flow rates of mobile phases. Three main types of flowmeter are employed. The most well-known is the vertical tube in which the frictional force of the mobile phase is used to lift a ball or small cone in a length of calibrated glass tubing (Figure 22). The flow rate in this device can be read directly from the scale on the tube. It gives very accurate values for gas flow, but for liquid flow is subject to errors of ± 5% and has to be separately calibrated for each liquid employed.

An alternative gas flowmeter measures the flow pressure difference on a manometer as the gas passes through a capillary. This type of instrument really needs to be individually calibrated for each carrier gas used.

The third type of flowmeter commonly employed for gases is the bubble flowmeter. It consists very simply of a vertical calibrated tube with a small reservoir of soap solution at the bottom. The carrier gas passes into the tube just above the soap and when the rubber reservoir is squeezed manually a bubble is formed which is carried up the tube. The rate of flow is determined by measuring the time the bubble takes to travel a fixed distance along the tube.

Flow programming (Pressure programming). A technique employed in both GC and HPLC to increase the rate of elution of slow-moving components. It involves progressively increasing the rate of flow of the mobile phase by an increase in the pressure. In many cases flow programming is not carried out until after an initial **isorheic** period. The flow programming may be linear,[179] with a constant increase in the flow rate, or stepwise, with periodic increases in flow rate. The technique of flow programming in HPLC cannot readily be used with **bulk property detectors** but is satisfactory with **solute property detectors** such as the **ultraviolet absorption detector**.

Flow rate (F_c). Mobile-phase flow rates are usually given in cm^3 min^{-1}, and for GC the column temperature and outlet pressure should be specified. In HPLC solvent flow rates vary between 0.25 and 5 cm^3 min^{-1}, with applied pressures from 200 to 6000 pounds per square inch (13.3–400 kg cm^{-2}); for GC gas flow rates are between 0.1 and 50 cm^3 min^{-1} at applied pressures of 12–65 pounds per square inch (0.8–4.6 kg cm^{-2}).

Although mobile phase flow rates are frequently determined empirically, the optimum flow rates are best obtained by plotting values for the height equivalent to a **theoretical plate**, calculated from the appropriate form of the **van Deemter equation**, against the mean gas velocity (*see* Figure 29, page 95). In practice the optimum conditions are not always applied as speed of operation may be of greater importance than a high degree of separation.

Flow splitters. *See* **Effluent splitter** and **Sample splitter**.

Fluorescent indicators. One of the difficulties with both PC and TLC is in the viewing of the separated components, as the majority of compounds are colourless unless specially treated. One method of overcoming this problem is to spray the chromatogram with a fluorescent dye. When placed under ultraviolet light any ultraviolet-absorbing compounds will then appear as dark spots within the fluorescing background. In the case of TLC the necessity of spraying can be overcome by incorporating a small quantity of fluorescent dye in the adsorbent when making up the plates. Such TLC adsorbents are commercially available premixed containing up to 1% by weight of the dye. Fluorescent indicators that have been used for this purpose include fluorescein, 2,7-dichlorofluorescein, and barium diphenylamine sulphonate.

Fluorimetric detector. A device by which fluorescent compounds can be detected in HPLC by passing the effluent through a cell irradiated with ultraviolet light and measuring any resulting fluorescent radiation.[180] This type of detector is quite specific, being limited in application to selected groups of substances such a families of drugs. Its sensitivity is about 10^{-9} g cm^{-3}. Modern microcells with sample volumes of 20 μL have been designed to fit direct into cell holders of standard spectrofluorimeters to obviate the need to obtain separate fluorimetric detectors.

Flying-spot densitometer. *See* **Densitometer**.

Foils. Very high quality thin layers on aluminium foils and plastic sheets produced as commercially available plates for TLC. The high standards of production enable very reproducible results to be obtained with the added advantages that the sheets can be cut to smaller sizes and shapes and are easily written upon as well as being less vulnerable to mechanical damage. Layers of thicknesses from 250 μm to 1 mm are available both in sheets and rolls.

Fractional development. A process used in PC or TLC when substances are only poorly separated by one solvent system. The chromatograms are run with solvents of increasing polarity, having being completely dried before each subsequent solvent is applied. The technique is used to achieve progressive separation of closely related substances.[181]

Fractionation range. The very nature of **gel chromatography** means that the gels employed differ in their abilities to separate chemical compounds. This difference is shown by the fractionation range to which each can be applied, as it corresponds to the upper and lower molecular weight values between which useful separations can be achieved. The upper value corresponds to the **exclusion limit** for that gel. Fractionation ranges may be only a few hundred (e.g. 100–5000) in the case of dense **dextran gels**, or extend to many thousands (e.g. 1000–150 000) for less dense gels (*see* Figure 43, page 169).

Fraction collectors. Manual collection of fractions from chromatographic columns can be very tedious and collection is best carried out using automatic fraction collectors. For conventional columns acting under gravitational forces the fraction collector consists of a large turntable with a series of test tubes arranged in a continuous ring around the circumference. As the eluate flows from the column it is conducted to one of the tubes, and after a predetermined volume has been delivered (in a set period of time) the table automatically rotates the following tube into position to receive a corresponding aliquot. The individual fractions may then be studied at leisure (Figure 23). For general application such instruments should be widely adjustable with timings between 5 s and 10 min and variable drop rates.

Similar devices are used for preparative GC, only in this case a small number of collecting tubes almost completely immersed in a freezing bath is used. The turntable holding the tubes is programmed so that after each automatic sample injection has been made on the column the tubes are positioned in turn to receive the appropriate solute. By this means each tube collects the same solute fraction from each separative cycle.

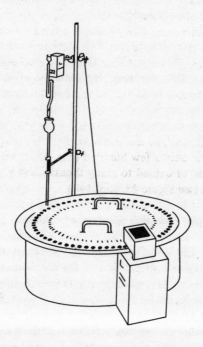

Figure 23. Fraction collector for column chromatography

Frequency difference radio-frequency detector. A detector which uses the change in dielectric constant of a **plasma** as a means of detecting solutes in the helium carrier gas flow.[182,183] The cell is a modified capacitor of small internal volume (20 μL) forming part of a circuit designed to oscillate at 65 MHz beat against a corresponding oscillator to a zero beat frequency. The presence of a solute leads to a change in the capacitance and hence a change from the baseline conditions. The best response from the detector is obtained with an electrode distance of 0.03 cm and 390 V plate voltage. The limit of detection is less than 10 p.p.m. for most substances studied, with a sensitivity of 1×10^{-8}

g s^{-1}. It is mainly suitable for the determination of inorganic gases, such as nitrogen, carbon dioxide, nitrous oxide and sulphur dioxide. *See also* **Radio-frequency detector.**

Fresnel refractometer detector. *See* **Refractive index detector.**

Freundlich's adsorption isotherm. The variation of the amount of gas adsorbed per unit area of surface was expressed in terms of an empirical relationship by Freundlich:

$$m = kp^{1/n}$$

where m is the mass of gas adsorbed, p the area, and k and n are empirical constants. A plot of log m against log p should give a straight line with a slope equal to $1/n$ and an intercept on the ordinate corresponding to log k. *See also* **Langmuir adsorption isotherm.**

Frontal analysis. The least frequently applied of the three methods for running chromatograms. It consists of the continuous addition of the dissolved mixture to the column, with the result that only the least sorbed compound is obtained in a pure state (Figure 24) as it moves at the fastest rate. As more of the solution is added to the column the second component is eventually eluted with the fastest moving material, and ultimately, if the procedure is continued, the eluate is identical to the mixture added at the column head.

Frontal analysis is really a semipreparative procedure, but column conditions must be carefully selected to ensure that the required material is obtained well before other substances are eluted from the column.

Desalting techniques in gel chromatography are a form of frontal analysis, although the frontal analysis procedure is most easily applied in ion exchange columns. *See also* **Displacement chromatography** and **Elution chromatography**.

Fronting. A term used to refer to peaks on chromatograms in which part of the peak tapers in advance of the bulk of the material so that

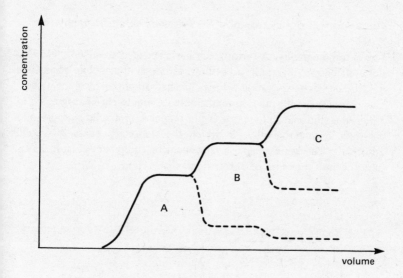

Figure 24. Frontal analysis

there is an asymmetric distribution with a leading foot. On PC and TLC the same type of effect gives rise to spots with streaks in front of the bulk of the substance. *See also* **Tailing**.

G

Gas adsorption chromatography. *See* **Gas–solid chromatography**.

Gas chromatography. A term which has been recommended[184] to describe all chromatographic procedures in which the mobile phase is a gas; the earlier term, vapour phase chromatography, is no longer used. Gas chromatography, therefore, includes gas–liquid chromatography, in which the stationary phase is a liquid retained on a solid support, and gas–solid chromatography, in which the stationary phase is a solid adsorbent. The same general arrangement of equipment (Figure 25) is used for both forms of gas chromatography.

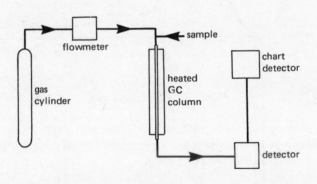

Figure 25. Block diagram for gas chromatography

Gas density balance. An instrument which responds to density changes in the gas stream, directly related to the molecular weights. Several forms of gas density balance have been developed since this type of detector was first introduced by Martin and James.[185] Improvements in instrumentation have led to a revival of interest in it as a detector system suitable for molecular-weight studies. The density changes are measured by means of a form of U-tube, one limb of which contains pure carrier gas and the other column effluent gas. An anemometer situated between the two limbs detects and measures any density difference (Figure 26). Using nitrogen or argon carrier gas the limit of detection is 0.4 μg, response is linear with concentration, and it can be used for molecular weight determinations[186] for values below 200 and for percentage composition studies.[187]

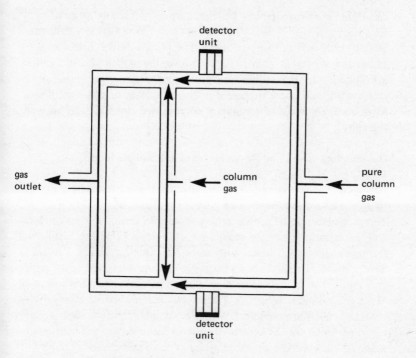

Figure 26. Diagram of a gas density balance

Gas-flow impedance detector. A detector based upon the principle that the pressure developed across a choke through which a gas flows is a function of the gas. As a result, at constant flow rates any pressure changes correspond to solutes being eluted in the gas stream.[188] A differential comparison of the flows of two gas streams, one the column gas and the other a reference flow, is made using a simple manometer. The detector is claimed to be at least as sensitive as the **katharometer**.

Gas hold-up volume (V_M). *See* **Hold-up volume**.

Gas–liquid chromatography. A technique arising from the suggestion by Martin and Synge[189] that the mobile phase for partition chromatography could just as easily be a gas. The procedure was introduced by James and Martin[190,191] in 1952. It involves the partitioning of a solute, introduced into the mobile gas phase, between the gas phase and a stationary liquid phase deposited upon the surface of either small solid particles or the walls of a **capillary column**. *See also* **Gas–solid chromatography**.

Gas sampling valves. *See* **By-pass injector** and **Sample loop**.

Gas–solid chromatography (Gas adsorption chromatography). Separation of mixtures by the use of a mobile gas phase and a solid adsorbent. The procedure can be shown to have developed from the work of Hesse and Tshachotin[192] on the separation of aliphatic acids by distillation through a silica gel column using carbon dioxide carrier gas. The major development of GSC as a separative procedure was made by Cremer and Prior[193] with their separation of acetylene-ethylene mixtures using silica gel columns, hydrogen carrier gas and a thermal conductivity detector (*see* **Katharometer**). With the development of special polymeric materials for the adsorbent column packings, gas–solid chromatography[194] has become almost as useful as **gas–liquid chromatography**.

Gas syringe. *See* **Syringes**.

Gas volume detector. *See* **Janak's method**.

Gaussian distribution (Normal distribution). The most common frequency distribution for continuous variables. The equation describing it gives the probability at a value x for a set of variables possessing a mean μ and a standard deviation σ:

$$P(x) = \frac{1}{(2\pi)^{1/2}\sigma} \exp \left\{ \frac{-(x-\mu)^2}{2\sigma^2} \right\}$$

The Gaussian curve shape is such that the width w_h at half-height $H/2$ is related to the peak height H by the equation

$$w_h = \frac{1.476}{H}$$

This is the ideal distribution of a component separated by a chromatographic process; it is symmetrical about its midpoint and the peak width is equivalent to 6σ. In practice the Gaussian shape is more easily obtained with small sample sizes than with large.[195] *See also* **Poisson distribution**.

GC. *See* **Gas chromatography**.

GC/IR interfaces. Although many attempts have been made to conduct GC effluents straight to infrared spectrometers it is only in recent years that real success has been achieved.[196] This is due to the development of Fourier transform techniques which have enabled detection in the nanogram range to be attained.[197] In such systems the eluate is divided into two streams, the first passing to a GC detector which, on producing a signal, switches on the interferometer as the other stream is passing through a 40 cm × 0.3 cm internal diameter flow cell fitted with KBr windows. The effective volume of such cells may be less than 1 cm^3 and the mixing of separated compounds is prevented by maintaining laminar flow. *See also* **LC/IR interfaces**.

GC/MS interfaces. About eight main procedures[198] have been devised to enable solutes from gas chromatographs to be passed straight to mass spectrometer ionization chambers. In all cases the major problem to be overcome is the removal of sufficient carrier gas to give a reasonable enrichment of the solute prior to entry into the spectrometer.

One of the earliest successful procedures was by Ryhage,[199] who used molecular separators[200] in series between the gas chromatograph and the mass spectrometer. This arrangement increased the sample-to-helium ratio by a factor of 100. Another procedure has been the use of membranes,[201] while the **Watson–Biemann interface**,[202,203] employing a sintered glass tube, has become widely used because of the ease with which it can be constructed.

Of major importance in all interfaces are (*a*) the enrichment factor, which is the relative increase in concentration of the compound after passing through the interface, and (*b*) the efficiency, the actual percentage of solute which passes to the ion source in the mass spectrometer. Modern interfaces produce enrichment factors of 100–10000 with efficiencies of 50–90%.

In the absence of these forms of interfacing it is necessary to collect the individual fractions from the gas chromatograph and apply them separately to the mass spectrometer. *See also* **LC/MS interfaces.**

Gel. The complete matrix used in **gel chromatography**, consisting of two component parts: the dispersed material (the solid) and the dispersing material (the solvent). The word is, however, frequently used to refer to the solid packing. A complete gel then consists of a continuous system in which the two components mutually penetrate and stabilize each other,[204] the structure comprising intertwined macromolecules and solvent molecules associated by a combination of hydrogen bonds and covalent bonds. *See also* **Aerogels; Agarose; Sephadex; Xerogels.**

Gel chromatography. A general inclusive term proposed[205] for all forms of **exclusion chromatography**. It is, however, more particularly used for those forms of chromatography which have previously been referred to as either gel permeation chromatography or gel filtration. It therefore refers to liquid separative procedures in which molecules are separated primarily according to differences in their molecular dimensions by their abilities to penetrate (Figure 27) a three-dimensional matrix formed from either polymeric organic materials, the **xerogels**, or inorganic glasses, the **aerogels**. Gel chromatography is of particular value for separating high molecular weight materials[206] such as proteins, and is used very widely in biochemistry. It may also be applied to the study of other polymeric materials such as waxes and plastics.[207]

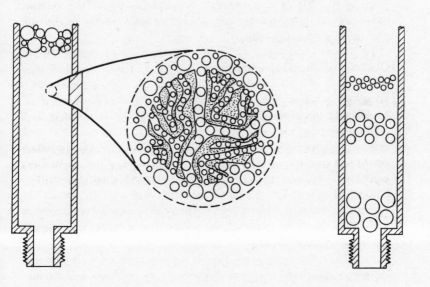

Figure 27. Gel chromatography

Gel filtration. *See* **Gel chromatography**.

Gel permeation chromatography. *See* **Gel chromatography**.

Glass aerogels. *See* **Aerogels**.

Glass beads. Although the use of glass beads is normally associated with conventional fractionation columns, they have also been used as solid supports[208] for the stationary phases in **gas–liquid chromatography**. Due to the nature of the glass surface, very thin coatings of stationary phases are possible: with the diesters of phthalic acid, such as **dinonyl phthalate**, a loading of less than 1% on 150–200 mesh size beads is used.

Specially etched borosilicate glass beads with controlled pore sizes are used as rigid glass **aerogels** in HPLC. The beads are within the size range of 40–200 μm with pore sizes between 40–500 Å. They have the advantages of being mechanically and chemically stable and of not swelling in different solvents.

GLC. *See* **Gas–liquid chromatography**.

Glueckauf equation. The first equation, introduced by Glueckauf,[209,210] to describe the separating efficiency of ion exchange columns by a continuous flow treatment in place of the discontinuous flow treatments employed by Mayer and Tompkins (*see* **Mayer–Tompkins theory**). It was proposed prior to the development of the **van Deemter equation**[211] and was originally used by Martin and Synge (*see* **Partition chromatography**). The Glueckauf equation was used to predict the number of theoretical plates required to achieve a particular separation and to calculate the shape of the solute elution curve. The curve is described by the equation

$$c = c_{max} \exp\{\frac{-n(\bar{\nu} - \nu)^2}{2\bar{\nu}\nu}\}$$

where ν is the volume of eluting solvent and $\bar{\nu}$ is the elution volume at the peak concentration c_{max}.

van Deemter *et al*[212] acknowledged that their equation was developed with Glueckauf's work in mind but their equation has been of much wider application in chromatography in general.

Golay columns. *See* **Capillary columns**; **PLOT columns**; **SCOT columns**; **WCOT columns**.

Golay equation. A modified form of the **van Deemter equation**[213] introduced by Golay[214] to describe fully the separating efficiency of **capillary columns** in terms of the **height equivalent to a theoretical plate**. In its shortened form this equation is presented as

$$h = \frac{B}{\bar{u}} + C_L \bar{u} + C_G \bar{u}$$

where C_L and C_G are the coefficients of mass transfer in the liquid (stationary phase) and gas (mobile) phase respectively — see below.

It should be noted that this equation does not include the first term, the **eddy diffusion term** of the van Deemter equation as there is only one path along the capillary column that the gas can follow, hence $A = 0$.

In the above equation C_L and C_G are obtained as follows:

$$C_L = \frac{2k}{3(1+k)^2} \frac{d_f^2}{D_L} \qquad C_G = \frac{1 + 6k + 11k^2}{24(1+k)^2} \frac{r_o^2}{D_G}$$

Gradient device. Two forms of gradient device are available for HPLC. In the exponential device the solvents are premixed before passing to the pump, the second solvent (usually the more polar of the two) being added to the primary solvent with rapid mixing while the pump is drawing solvent from the primary solvent reservoir. The rate of mixing can be controlled to give an exponential, linear or stepwise gradient.

Gradient elution is, however, easiest with two separate pumps jointly controlled to pump the two solvents at a predetermined changing rate into a small highly efficient mixing chamber from which the resulting mixture is passed directly to the column. The whole system is arranged to ensure that any delay between mixing and reaching the column is kept to a minimum.

Gradient elution. A process that although used on conventional liquid–liquid chromatograms for many years has really grown in application since the development of HPLC.[215] The polarity of the initial solvent is progressively changed by the continuous programmed addition of increasing quantities of a second solvent. The gradient elution may be carried out by initially running for an **isocratic** period with the first solvent alone before carrying out the gradient addition. The actual gradient used may be linear, exponential or stepwise as required, depending upon the degree of control possible with the instrumentation. This form of solvent programming in HPLC has a similar effect to the **temperature programming** in GC; the increasing amount of polar solvent causes the more strongly retained solutes to move more rapidly, enabling the chromatogram to be completed in less time than is required for an isocratic elution using just a single solvent. Gradient elution is used extensively in **reversed phase systems**.[216]

GSC. *See* **Gas–solid chromatography**.

H

Headspace analysis. A sampling technique for GC[217,218] that has developed extensively because of the sampling problems[219] previously associated with direct injection of blood and urine samples for the quantitative determination of ethanol. It has also found application in an extensive range of analyses, including sewerage, vapours released from fruits, volatile compounds in polymers, and biological fluids.[220]

The procedure involves taking an aliquot of the vapour above the sample (the headspace) contained in a closed vessel. It is based upon the idea that at equilibrium under controlled conditions[221] the atmosphere will be representative of the sample (although proportions of components will depend upon partition between the sample and the air above it). The actual sampling procedure may be carried out in a variety of ways, but the most common method is to have the substance to be analysed contained in a small sample container fitted with a rubber septum cap. After thermostating the container an aliquot of the headspace is taken by a gas syringe (Figure 28) and injected onto the GC column. Where large numbers of similar samples are to be analysed it is common to use highly sophisticated automatic headspace sampling procedures. Even so a wide range of possible errors have to be guarded against, particularly absorption of vapour by the septum.[222,223]

Heat of adsorption detector. *See* **Thermal adsorption detector.**

Height equivalent to a theoretical plate (*h*). The height of a layer on a chromatographic column such that the mean concentration of solute in the stationary phase is in equilibrium with the solute in the mobile phase at that layer. The relationship between *h* and the mean gas velocity is shown in Figure 29. The theoretical basis for this has been developed in the **van Deemter equation**[224] and the various extensions for modified column conditions, for example the **Golay equation.**

For any particular column the value of *h* is obtained by dividing the length of the column, *L*, by the number of plates, *n*:

$$h = \frac{L}{n}$$

n is obtained from any peak on the chromatogram by the relationships

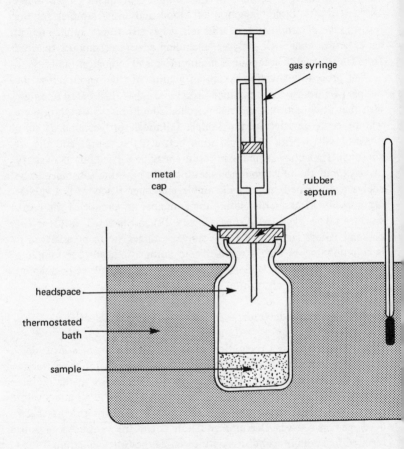

Figure 28. Headspace analysis

$$n = 16 \left(\frac{d_R}{w_b}\right)^2 = 5.54 \left(\frac{d_R}{W_h}\right)^2$$

At low gas flow rates h is determined mainly by the molecular diffusion term of the van Deemter equation, but at high gas velocities the contribution from this term is small and the major factor is the resistance to mass transfer. H is now recommended for use as the symbol for the

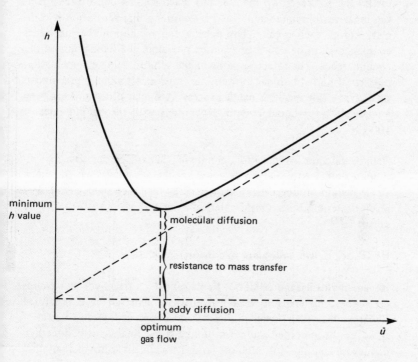

Figure 29. Relationship between the height equivalent to a theoretical plate and the flow rate

height equivalent to an effective theoretical plate (HEETP), based upon column performance and effective resolution. *See also* **Column performance** and **Efficiency of a column.**

Helical flow columns. Continuous chromatography by using helical flow was a suggestion of Martin[225] that was later developed by other research workers.[226] The system consists of a series of fixed vertical columns (100 columns, each 1.2 m long and 6 mm internal diameter) arranged in a circle to rotate at from 1 to 50 revolutions per hour. During rotation they pass in turn under a feed inlet and carrier gas then carries the solutes down the columns. Each column constitutes a conventional gas chromatograph, and by ensuring that separation is complete before each column has done a full revolution the individual components can be collected from each column at a fixed point on the circumference. The effective paths of the solutes, being a combination of vertical and rotational motion, are a helix. Although a preparative method,[227] this is still a **batch process.** A similar procedure has been employed[228] for liquid column separations with throughput rates of 100 g h^{-1}.

Helium detector. A detector, employing helium carrier gas, that although similar to the **argon detector** has the advantage that the metastable helium produced has a higher energy than metastable argon. As a result it responds to most materials[229] and is claimed to be highly sensitive.[230]

HETP. *See* **Height equivalent to a theoretical plate.**

Hexamethyldisilazane (HMDS). Replacement of free hydroxyl groups on support materials to deactivate them is achieved by a variety of **silylation** reagents. Hexamethyldisilazane is one of the most useful of these as the modified supports, unlike those treated with **dimethyldichlorosilane,** do not require further treatment with methanol.[231] The deactivation treatment is carried out by refluxing the support in a petroleum ether solution of HMDS, the following type of reaction occurring on the active support sites:

$$
\begin{array}{ccc}
\begin{array}{c} | \\ -\text{Si}-\text{OH} \\ | \\ \text{O} \\ | \\ -\text{Si}-\text{OH} \\ | \end{array}
&
+
\begin{array}{c} \text{Si}(\text{CH}_3)_3 \\ | \\ \text{NH} \\ | \\ \text{Si}(\text{CH}_3)_3 \end{array}
\longrightarrow
&
\begin{array}{c} | \\ -\text{Si}-\text{O}-\text{Si}(\text{CH}_3)_3 \\ | \\ \text{O} \\ | \\ -\text{Si}-\text{O}-\text{Si}(\text{CH}_3)_3 \\ | \end{array}
\end{array}
$$

The reagent has also been used to prepare volatile derivatives, suitable for GC separation, from nonvolatile starting materials.

High performance liquid chromatography (HPLC). A procedure, usually referred to as HPLC, whose development has revolutionized liquid chromatography and is applicable to far more compounds than is GC. All forms of column chromatography involving a liquid mobile phase can be extended to HPLC; liquid–liquid partition, liquid–solid adsorption, gel chromatography and ion exchange processes have all been successfully carried out.[232] The overall procedure[233] is characterized by the use of small particle sizes (down to 5 μm), narrow bore columns (1–7 mm internal diameter) and high column inlet pressures (up to 600 atmospheres) to achieve separations in short periods of time from a few minutes to one hour.

Equipment for HPLC (Figure 30) is usually constructed in stainless steel and requires special pumps to attain the high pressures applied. Specially strengthened glass columns can be used up to pressures of 65 atmospheres ($\approx$ 1000 psi).

High performance thin layer chromatography. A term coined for TLC carried out on specially made high quality thin layer plates capable of achieving high resolution between compounds with rapid running times and often over short distances. In some instances separation distances are reduced by 75% and running times by 80% compared with conventional TLC, while detection limits are improved by a factor of 10. This improvement has been achieved by plate manufacturers maintaining very tight control over production and the design of chromatographic chambers in which atmospheres over the chromatogram are totally constant with vapour saturation achieved throughout.

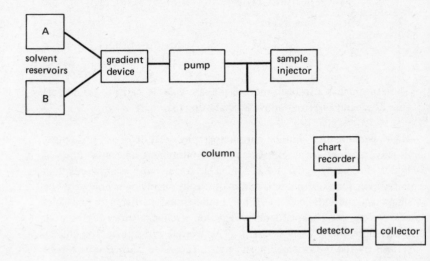

Figure 30. Block diagram of high performance liquid chromatography

High pressure liquid chromatography. *See* **High performance liquid chromatography.**

High speed liquid chromatography. *See* **High performance liquid chromatography.**

High voltage zone electrophoresis. Separation of low molecular weight substances, such as inorganic ions, phenols and carbohydrates, by zone **electrophoresis** is best carried out by high voltage procedures. Voltages up to 10 kV have been used. Such methods are more rapid than **low voltage zone electrophoresis** but cooling of the system is essential. Removal of the excess heat from the electrophoresis strip is best achieved by either **enclosed strip** or **immersed strip electrophoresis.**[234]

HMDS. *See* **Hexamethyldisilazane**.

Hold-up volume (V_M). The volume of eluent required to elute a component moving with the part of the mobile phase into which it is injected and undergoing no delay due to interaction with the stationary phase. This includes any dead volume arising from the injection and detection.

Horizontal chromatography. A procedure, involving the use of a shallow tray, which is employed when suitable equipment does not exist for carrying out PC using ascending or descending methods. As illustrated in Figure 31, the paper is supported on a series of glass rods above the solvent with one edge of the paper held in the solvent.

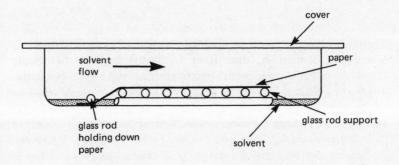

Figure 31. Horizontal chromatography

Hot wire detector. *See* **Katharometer**.

HPLC. *See* **High performance liquid chromatography**.

hR_F value. *See* **R_F value**.

HSLC. Abbreviation for high speed liquid chromatography. *See* **High performance liquid chromatography**.

Hydrogen bonding. A weak bond frequently formed by hydrogen atoms in molecules in addition to the main covalent molecular bond. Typical examples of hydrogen bonding are between water molecules, with carboxylic acids, and with amides:

$$H—O—H \cdots O—H$$
$$\qquad\qquad\quad | $$
$$\qquad\qquad\quad H$$

$$R—C \overset{O \cdots H—O}{\underset{O—H \cdots O}{}} C—R$$

$$R—C \overset{O \cdots H—N}{\underset{N—H \cdots O}{}} C—R$$

The dotted line represents the hydrogen bond in all cases. This form of bonding, or association, arises from a strongly electronegative atom (O, N, F) participating in partial electron sharing with a hydrogen atom attached to nitrogen or oxygen. In many cases these structures can also be described in resonance terms in which the stable form involves a full sharing of the protons between two electronegative atoms.

In chromatography, hydrogen bonding between solutes and stationary phases can be a major cause of peak tailing and poorly defined separations. It is for this reason that **deactivation** of some solid support materials is necessary.

Hydrogen flame detector. *See* **Flame thermocouple detector.**

Hydrogen microflare detector. *See* **Flame thermocouple detector.**

Hydrophilic. A word, literally meaning water loving, which is used in connection with substances which have an affinity for water and are capable of associating with water molecules through the formation of hydrogen bonds (*see* **Hydrogen bonding**).

Hydrophobic. A word, literally meaning water hating, which is used to describe those substances which have little affinity for water molecules and will repel them. Compounds in this group usually possess few functional groups and have essentially a hydrocarbon or aromatic structure.

I

Immersed strip electrophoresis. A method of **high voltage zone electrophoresis** in which the two ends of the strip are dipped in buffer solution while the main body of the strip is covered by an insulating cooling liquid. By this means buffer solution evaporation losses and **capillary flow** are minimized. The most suitable liquid used is silicone oil, although toluene and carbon tetrachloride have been used.[235] *See also* **Enclosed strip electrophoresis.**

Immunoelectrophoresis. A method involving a combination of electrophoresis and specific immunochemical reactions[236] by which proteins can be detected and identified. The procedure is carried out on cellulose acetate strips, upon which the serum proteins are initially separated by conventional zone **electrophoresis.** The celluluse acetate strip is then placed on a glass plate covered with agar gel[237] and a strip of filter paper soaked in antiserum is placed alongside touching the edge of the cellulose acetate strip. This arrangement is maintained at 37 °C in a moist atmosphere for several days during which diffusion of antigens (serum proteins) and antiserum occur. Where the two species meet, arcs of precipitate form.[238] These can be directly photographed on the gel after the strips have been removed, and may be stained by conventional means.

Infrared detector. A wide range of commercial infrared detectors for use with HPLC have been developed with flow-through cell volumes from 5–200 μL and optical path lengths of 1–3 mm. Cell windows in the normal infrared-transparent materials (sodium chloride, calcium fluoride, KRS-5, etc.) are available and full wavelength scanning between 2.5 to 14 μm is possible. These instruments are particularly suitable for detecting chemically related substances by scanning at a selected or narrow wavelength band. They have the great advantages of being insensitive to temperature and flow-rate changes.[239]

Injection port. A device that is necessary in all GC and HPLC systems

102

in which syringe injection is employed. It joins the top of the column to the mobile phase supply line, and has a third outlet through a small orifice covered by a self-sealing **septum**. Injection of samples is done by inserting the hypodermic syringe containing the sample through this septum so that the point of the needle is immediately above the head of the column. Design of the injection port must be such that it can be maintained at a uniform temperature and will possess a very small **dead volume**.

Inlet splitter. *See* **Sample splitter.**

Inlet systems. The simplest form of sample inlet system is the **injection port** for injection by hypodermic syringe. Other more complicated inlet systems are frequently required for use with automatic sampling devices, pyrolysers (*see* **Pyrolysis gas chromatography**) and gas **by-pass injector** valves. In some instances the additional piece of apparatus is designed to screw onto the standard injection port in place of the septum-retaining screw.

It is impossible to design an inlet system without a finite volume, but its contribution to the overall dead volume of the system must be kept to a minimum. The design of inlet systems has come in for much criticism[240] because of inadequate attention to the effects of diffusion and the capacity of the chamber.

Inner volume (V_I) (Internal volume). In gel chromatography, the volume of the solvent which has penetrated the gel matrix. The value of V_I is related to the porosity of the gel by the equation

$$V_I = \frac{aS_r}{\rho}$$

where a is the weight of the dry gel, S_r is the **solvent regain** and ρ is the specific gravity of the solvent.

Values of V_I are difficult to measure accurately and often have to be determined indirectly. *See also* **Total volume.**

Instruments for elemental analysis. *See* **C, H, N analysers.**

Integral detectors. According to Huber,[241] a detector which responds to an input signal which is proportional to the overall amount of sample which has been eluted from the column. Mathematically this may be expressed as

$$X = a_i Q$$

The chart record from such detectors shows a cumulative increase in the value of the measured property as the quantity of solute in the mobile phase passes through the detector, and consists of a series of ascending steps (Figure 32) the height of each step corresponding to the amount of each solute. *See also* **Differential detector.**

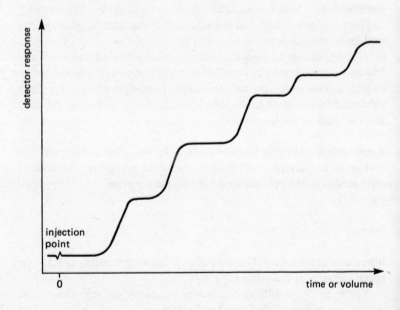

Figure 32. Integral detector signal

Integration. In the absence of an **integral detector**, integration of chromatographic peaks is carried out by manual methods (**cut and weigh**, measurement of peak heights, use of a **planimeter, triangulation**), by mechanical means with a disc integrator, or electronically with a computer and tape or digital readout. Of these various methods the electronic and mechanical procedures are the most accurate, with the former being the most expensive. The least accurate are the cut and weigh and the measurement of peak heights.

Interfaces. *See* **GC/IR interfaces; GC/MS interfaces; LC/IR interfaces.**

Internal standards. Means by which precision and accuracy in quantitative analysis are greatly improved. They are employed extensively in GC determinations for the ethanol content of blood and urine[242,243] and for the presence of unreacted monomers in polymeric materials.[244]

The procedure involves adding a fixed amount of an internal standard both to a series of increasing concentrations of reference sample and to the unknown concentration.[245] The ratio of the areas of the reference concentrations and the areas of the internal standard is plotted against the known concentration of the reference samples. This ideally gives a straight-line calibration plot (Figure 33) and the unknown sample concentration can be established by determining the position of the corresponding ratio of the sample area and internal standard on the calibration line.

The substance employed for the internal standard should be chemically similar to the substance to be quantitatively determined and have a retention time fairly close to it. *See also* **Standard addition.**

Internal volume. *See* **Inner volume.**

Interstitial fraction (ϵ_I). The interstitial volume per unit volume of the packed column. It is represented mathematically as

$$\epsilon_I = \frac{V_I}{X}$$

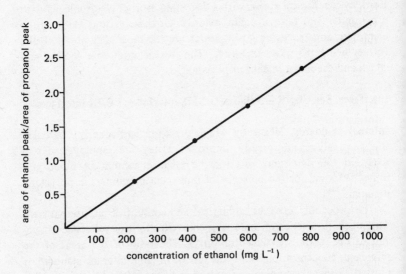

Figure 33. Calibration curve for the analysis of ethanol in aqueous
solution using propanol as an internal standard

Interstitial velocity (u). The mobile phase linear velocity averaged over
the total cross section:

$$u = \frac{F}{\epsilon_I}$$

For a carrier gas the mean interstitial velocity $\bar{u}$ is a quantity obtained
after correcting for the pressure gradient:

$$\bar{u} = \frac{Fj}{\epsilon_I}$$

Interstitial volume (Void volume) (V_I and V_G). In all forms of column chromatography, the total volume of mobile phase within the length of the column. It corresponds to the retention time for air in GC and hence to the **dead volume** of the chromatograph. In GC the interstitial volume is corrected to allow for the pressure reduction at the outlet, this corrected volume having the symbol V_G:

$$V_G = \frac{V_I}{j}$$

Intrastitial volume. *See* **Stationary liquid volume.**

Ion chromatography. A term which has been used[246] to refer to all techniques used for separating and measuring ions with pK_a values < 7. It is the chromatography of ions usually employing ion exchange columns and electrical conductivity cells.

Ion exchange chromatography. An inclusive term for all chromatographic separations of ionic substances carried out by the use of an insoluble **ion exchanger** as the stationary phase.[247] The passage of the ionic solution over and through the exchanger occurs with ions from solution exchanging for those already on the solid matrix.[248,249] The earliest ion exchange separations were carried out on naturally occurring **zeolites** before contemporary chromatographic and ion exchange procedures had been developed.[250]

Nowadays two main forms of ion exchanger are commonly used: **cation exchangers** and **anion exchangers**.

Ion exchange isotherm. A measure of the distribution of an ion between the ion exchanger and the surrounding medium. It is expressed as the concentration of the ion on the exchanger relative to its concentration in the external solution under defined conditions.

Ion exchange membranes. **Anion** and **cation exchangers** in the form of flexible sheets about 0.05 cm thick, available for use in electrolytic cells for **electro-osmosis** and **dialysis**. The strongest membranes are those made by incorporating the ion exchange resin into plastic sheets of polythene or polyvinyl chloride. Such membranes are semipermeable, permitting only the passage of the same charged ions as those they will exchange.[251]

One great advantage that membranes possess over ion exchange columns is that the total resin is more readily accessible to the solution at all times. In electrodialysis equipment alternate sheets of cation exchange membranes and anion exchange membranes are used to achieve the desalting effect.

Ion exchange paper chromatography. Ordinary paper is a weak cation exchanger and this capacity can be improved by the incorporation of polyelectrolytes.[252] Similar papers can be made by adding finely ground **ion exchange resins** to pulped filter paper and redrying to give thin sheets.[253] Such papers are useful in the separation of closely related compounds like the amino acids and barbituric acid derivatives. Qualitative separations of metal ions are easily carried out on a small scale using ion exchange papers, and they have the advantage that R_F values can be determined. The retention of metal ions on ion exchange papers is not only due to the ion exchange process: it has been shown[254] that complex formation, adsorption and partition all play a part.

Ion exchanger. An insoluble solid or liquid polyelectrolyte possessing labile ions which are able to participate in exchange reactions with ions in the surrounding solution without altering the general physical nature of the ion exchanger. For practical purposes the desirable characteristics of an ion exchanger are that it should provide an insoluble chemically stable stationary phase possessing mechanical strength in a particle size able to flow reasonably easily for column packing.

The first use of ion exchangers was more than a century ago,[255] and until the introduction of the synthetic **ion exchange resins**[256] in 1935 the inorganic naturally occurring **zeolites** were used for ion exchange studies and water softening. *See also* **Liquid ion exchangers.**

Ion exchange resins. Synthetic organic ion exchange materials prepared by the incorporation of appropriate functional groups into polymeric structures. The structure may be based upon polystyrene and divinylbenzene, phenol and formaldehyde, or similar stable polymeric chemical structures. To function as a **cation exchanger** the resin is substituted with either sulphonic acid groups (for strong cation exchangers) or carboxylic acid groups (for weak cation exchangers). **Anion exchangers** are obtained by introducing primary and secondary amino groups into the polymeric lattice.

The first synthetic ion exchangers of this type were prepared[257] in 1935 using a phenol–formaldehyde polymer for the cation exchanger and a phenylenediamine–formaldehyde polymer for the anion exchanger. More recent advances in this field include the development of **chelating resins** and **amphoteric ion exchange resins**.

Ion exclusion. The process[258] by which separation of ionized substances from aqueous solutions which also contain slightly ionized and non-sorbed substances can be achieved on **ion exchange resins**. This is possible due to the **Donnan membrane potential** by which ionic solutes exist at a higher concentration in the solution than in the resin whereas nonionic solutes are distributed uniformly between the solvent in the column and that within the resin matrix. As a result the ionic solute moves faster down the column than does the nonpolar (usually low molecular weight organic) compound.

Ionization cross-section detector. *See* **Cross-section detector**.

Ionization detectors. A family of related detectors which is based upon changes in the conductivity of electricity through gases arising as a result of the presence of charged particles and ions in the gas.[259] These ions may be created by combustion (as in the **flame ionization detector**), transfer of energy (in the **argon detector**), or by β ionization (in the **cross-section** and **electron capture detectors**).

The first detector of this series that was developed was the **cross-section detector** although the most widely used is the **flame ionization detector**. Their great popularity rests in their wide applicability and high sensitivity.

Ionization gauge detector. A detector system[260] which is formed from the combination of a thermionic ionization gauge and a mass analyser. The solute is ionized in the ionization gauge, partially collected (to give an electronic signal) and partially passed to the mass spectrometer. As each peak is detected on the ionization gauge a simultaneous determination of the molecular weight is made on the mass spectrometer. A specially designed collector electrode is employed in the ionization gauge ensuring that ions are only partially collected.

Ionogenic groups. The fixed groups in **ion exchangers** which are ionizable and can dissociate into mobile and fixed ions.

Ionophoresis. A name suggested by Martin and Synge[261] for the migration of relatively small ions under electrophoresis conditions in order to differentiate between this and the **electrophoresis** of proteins and macromolecules. It has, however, never been universally accepted and has dropped into disuse.

Ion pairing. **Reversed phase systems** are not greatly successful in separating strongly ionic compounds because of ionization effects. In weakly ionic compounds this is suppressed by suitable buffering, but in other cases it is necessary to combine the strongly ionic species with a counter-ion under appropriate pH conditions and this procedure is known as ion pairing.[262] Carboxylic acids, for example, may be treated with the strongly ionic tetrabutyl ammonium phorphate giving the following ionic reaction:

$$RCOO^- + (Bu)_4N^+ \longrightarrow RCOO^-N^+(Bu)_4$$
$$\text{counter-ion} \qquad\qquad \text{ion pair}$$

The ion pair is then able to participate in the normal HPLC processes.[263] However, controversy still exists concerning the mechanisms operating in ion-pair operation.

The method has been of particular value in the separation of drugs, antibiotics and vitamins. High resolution separation of sulphonic acids and dyestuff intermediates may be obtained by using cetyltrimethyl ammonium bromide (a detergent) at a 1% concentration in propanol-water solvent on HPLC;[264] this has been termed soap chromatography.

Ion retardation. A term which is used particularly to refer to the use of **amphoteric ion exchange resins**[265] in column procedures in which the combined effects of the cation- and anion-exchanging groups are employed to achieve the separation of an electrolyte (such as sodium chloride) from a liquid or solvent that is water soluble.

The electrolyte, in solution, is put on the ion exchange column and eluted with water. The electrolyte is preferentially retarded due to its ionic association with the resin, and thus the nonelectrolyte liquid is eluted with the water.[266]

Isocratic. A term used to refer to the continuous running of HPLC with a single solvent mixture. This method of operation can frequently be very time-consuming if solutes are strongly sorbed on the column and it is common to follow an initial period of isocratic operation with a **gradient elution** to speed up the rate of elution.

Isoelectric focusing. A procedure[267] which is based on the fact that if a pH gradient can be established between the cathode and anode in an **electrophoresis** system, then **zwitterions** will distribute themselves between the electrodes according to their **isoelectric points**. The growth of isoelectric focusing has depended upon the development of natural pH gradients from a series of **ampholytes**. The gradients usually cover a range of about 2 pH units. Isoelectric focusing is employed particularly for the separation of proteins[268] with which resolution between proteins differing by only 0.02 pH units can be achieved. One of the great problems in the procedure is the tendency of the proteins to precipitate out once there is any great concentration at the isoelectric point and this is minimized by using very low column loadings.[269]

Isoelectric point. Doubly charged ions such as amino acids all have a definite pH at which they will not migrate to either electrode under electrophoretic conditions. This pH value is known as the isoelectric point for the particular **ampholyte** and is characteristic for that substance. It is also the pH at which the substance has its lowest solubility, and in the case of proteins corresponds to that at which the highest osmotic pressure and viscosity for the solution occur.

Isorheic. In chromatography, a term used to refer to any period of constant flow of the mobile phase. An isorheic period of operation is commonly carried out before a period of **flow programming** and the chromatogram is often terminated with a final isorheic period at a faster flow rate.

Isotachophoresis. A development from **electrophoresis** in which conditions are established such that all ions of identical sign will travel at the same velocity.[270] The system depends upon having two like charged ions, one of which will normally move faster than the ions to be separated and one which will move slower. When a current is passed through the solution the fastest moving ion moves rapidly towards the appropriate electrode but cannot move ahead of the fastest moving ion in the sample being separated as this would produce a discontinuity in the conductivity of the solution. Similarly, no gap is possible between adjacent ions and the overall result is that the ions are separated into discrete zones immediately behind each other moving at a speed dictated by the slowest moving ion.

Applications of isotachophoresis are becoming of increasing importance and include separations of large numbers of ions simultaneously, e.g. 33 different cations,[271] and 12 aliphatic acids.[272] For analytical purposes such separations are best carried out in a capillary tube.[273]

Isothermal–isobaric chromatography. Chromatography carried out under conditions in which both the temperature and the rate of flow of mobile phase have been kept constant throughout the separation.

Isothermal operation. Any chromatography carried out under constant temperature conditions, including that at elevated temperatures in which the column is thermostatically controlled in an oven.

J

James–Martin gas density balance. *See* **Gas density balance.**

Janak's method (Gas volume detector). A procedure, introduced by Janak,[274] in which carbon dioxide is used as the carrier gas and emerges from the GC column to be absorbed in concentrated potassium hydroxide solution. Any unabsorbed gases pass through the solution and are collected in a gas microburette above the column outlet. Since the volume collected increases as the various solutes are eluted from the column, this device represents an **integral detector**. The procedure has been of particular value in the analysis and control of hydrocarbons in refinery gas streams.[275]

K

Katharometer (Thermal conductivity detector; Hot wire detector). One of the most widely employed detectors for GC, its popularity arising in particular from the ease with which it can be constructed within most competent laboratories.[276] Its operation is based upon the change in thermal conductivity of the carrier gas around a heated wire forming one arm of a **Wheatstone bridge** circuit. A change in thermal conductivity occurs whenever the carrier gas contains a solute; the change in heat loss causes a change in resistance of the wire which throws the bridge out of balance causing a flow of current which can be amplified and recorded. The change in thermal conductivity, and hence resistance, is compared with a reference resistance arm through which pure carrier gas is flowing (Figure 34).

Tungsten-rhenium filaments operating at 400 °C are commonly employed with an inert carrier gas such as helium, nitrogen or argon at a flow rate of 3 cm^3 s^{-1}. In conventional katharometers the sample cell volume is 0.1 cm^3, with micro versions available at 0.01 cm^3. Modern designs of the detector[277,278] are based upon the use of thermistors rather than wire filaments and cell design and shape are of major importance for use with open-tubular columns. Sensitivity values of 5 × 10^3 mV cm^3 mg^{-1} have been claimed with detection limits of 10^{-6} mol, so that katharometers are not as sensitive as flame ionization detectors. Despite these limitations their great attribute is that they are universal detectors.

K_{av}. A figure indicating that portion of the gel medium which is available to a particular substance for participation in the gel chromatographic process.[279] It is related to the other gel chromatographic parameters by the equation

$$K_{av} = \frac{V_e - V_o}{V_t - V_o} = \frac{V_e - V_o}{V_I + V_S}$$

The K_{av} figure is normally used instead of the K_D (distribution constant)

114

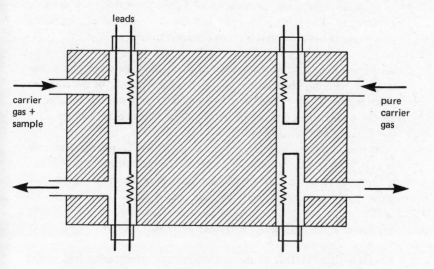

Figure 34. Thermal conductivity detector (katharometer)

value as the former is more easily determined, although the K_{av} figure is generally smaller, implying that a less than theoretical portion of the gel medium is available for that particular solute.

K_D. Strictly speaking, the **distribution constant** for a substance on a particular gel chromatography column, related to the other parameters by the equation

$$K_D = \frac{V_e - V_o}{V_I}$$

It has been stated[280] that the K_D value can be envisaged as representing the fraction of the total inner volume (V_I) to which the particular solute molecules have access. For totally excluded molecules $K_D = 0$, and for very small molecules able to penetrate fully into the available spaces $K_D \approx 1$.

In practice accurate estimations of V_I for this calculation are very difficult, so that instead of using the K_D value for calculating solute movement in the column it is more common to employ K_{av}.

Kieselguhr. One of the **diatomaceous earths** that are employed extensively in column chromatography and consist mainly of heat-treated remains of diatoms. Kieselguhr has a slightly pink colour, believed to be due to the small amount of iron oxide which becomes evident when the diatoms are heated to 900 °C.

It is used particularly as a support in partition chromatography, as it is only weakly adsorptive and possesses a large surface area (4-6 m² g⁻¹) and porosity, and is suitable for the separation of strongly hydrophilic compounds. It is easy to pack, has a pH 6.7 and will take a 5-30% stationary phase loading. For TLC it may have 5% added gypsum as a binder (kieselguhr G). *See also* **Celite**.

Kováts indices (I). One of the various systems of **retention indices** and related expressions that have been introduced as bases for reference and comparison of substances on gas chromatograms. Kováts indices were introduced to provide a retention index system[281,282] for substances relative to a scale obtained on the actual gas chromatograph employed. This is achieved by calculating the scale from the elution maxima of a homologous series of alkanes at a defined column temperature and stationary phase.

To use the system each alkane peak maximum is given a value of $100z$, such that for the series H_2, CH_4, C_2H_6, C_3H_8, C_4H_{10}, etc, the corresponding values are 0, 100, 200, 300, 400, etc, respectively. Using this scale the retention index (I) for any other substance is found by reference to its position between the two corresponding alkanes which would occur on either side of it on the chromatograms, and is calculated according to the following equation employing retention times for the three substances:

$$I_x = 100 \frac{\log t_x - \log t_z}{\log t_{z+1} - \log t_z} + Z$$

where t_x is the retention time for the unknown, and t_z and t_{z+1} are retention times for the two alkanes, Z being the number of carbon atoms in the alkane immediately prior to the substance studied.

It is usual to specify at which temperature the value has been determined by means of a suffix; for example I_{75} is a retention index determined at 75 °C. Retention indices calculated by this means are used extensively[283,284] and a nomogram has been devised to assist in their calculation.[285] *See also* **McReynolds' constants** and **Retention factor**.

L

Labyrinth factor. *See* **Molecular diffusion term**.

Langmuir adsorption isotherm. Langmuir considered the adsorption of gases on surfaces to consist of two actions in opposition to each other, one being the adsorption of molecules from the gaseous phase onto the adsorbent, and the other the desorption of the molecules back into the gaseous phase. His treatment also assumes that adsorption occurs on the solid with the formation of layers of solute no more than one molecule deep, i.e. monomolecular layers. This means that while the initial rate of condensation on a fresh surface is high the rate rapidly decreases as the available area becomes occupied with solute molecules. An equilibrium occurs when the rates of adsorption and desorption are equal. Under these conditions the mass of substance adsorbed per unit area of adsorbent is given by

$$m = \frac{ap}{1 + bp}$$

where a and b are constants characteristic for the system and are determined experimentally. *See also* **Freundlich's adsorption isotherm**.

Large bore open tube columns. *See* **LBOT columns**.

Layer. In chromatography a uniform distribution of an **adsorbent**, solid **support** or **gel** that has been spread on a flat surface such as a glass plate, aluminium foil or a sheet of plastic. A layer can be loosely packed or held firmly on the surface and can be considered to be a two-dimensional representation of a column.

LBOT columns. Large bore open tube columns, a relatively new development used to scale up results obtained on **capillary columns**. The open tube concept automatically leads to faster flow rates, and the various treatments of support coating and wall coating can be applied.

LC. *See* **Liquid chromatography.**

LC/IR interfaces. Although it is easy enough to develop small-capacity flow cells for infrared detection, the main problem in detecting solutes in LC eluates has been the rate of scanning of the infrared spectrum.[286] This needs to be rapid enough to differentiate between closely spaced species. The development of Fourier transform techniques has been the major advance in solving this problem so that with rapid almost instantaneous scanning, continuous monitoring of eluate is possible. *See also* **Infrared LC detector** and **GC/IR interfaces.**

LC/MS interfaces. Following the success of direct transference of GC eluates on to the mass spectrometer (*see* **GC/MS interfaces**), attempts have been made to achieve the same result with the eluates from HPLC.[287] The main problem in all cases has been the removal of the large quantity of solvent. An early approach to this involved passing the eluate into a hollow probe with a removable tip containing a gold gauze for the flash evaporation of the solvent before it entered the mass spectrometer ion source. In order to study individual fractions the chromatograph was operated on a stop-flow mode as each peak emerged.

An alternative treatment[288] has been to pass all the eluate on a continuously moving polymeric belt forming part of a direct inlet system. The solvent is rapidly removed by a strong infrared lamp and the sample flash evaporated within the ion source.[289] Another technique[290] is to concentrate the sample by allowing it to flow down an electrically heated wire and through a regulatory needle valve from which it is sprayed into the ion source. A twenty-fold increase in concentration is claimed for this method.

Ligand. Apart from its conventional usage in complexometry and with **chelating ion exchange** columns the word ligand is employed in **affinity chromatography** with reference to the biospecific material (the enzyme, antigen or hormone) which is coupled to the polymeric carrier to form the biospecific column.[291] The ligand must show a highly selective affinity for the macromolecule and possess a functional group suitable for attachment to the carrier.

Limit of detection. *See* **Lower limit of detectability**.

Linearity. For use in quantitative analysis it is essential that a detector should give a linear response with respect to solute concentration. No detector has perfect linearity over its full range of detection, and this range in any case varies with the nature of the solute. The linearity may be expressed either in terms of percentage deviation over a defined range or in terms of the exponent (x) of the equation

$$D = A_R c^x$$

where D is detector output, A_R is a constant (the response factor), and c is solute concentration in the mobile phase. For truly linear detectors $x = 1$, but in practice for all detectors x only equals unity over limited ranges and a detector is accepted as giving a linear response where values of x fall between 0.98 and 1.02.

One of the best detectors for quantitative analysis, from the point of view of an extensive linear range, is the **flame ionization detector,** whereas the **electron capture detector** has a much shorter linear range.[292,293]

Linear range. *See* **Linearity**.

Linear temperature programming. *See* **Temperature programming**.

Line oven. *See* **Ring-oven technique**.

Liquid chromatography. Any chromatographic system involving the use of a liquid mobile phase, including liquid–solid chromatography, liquid–liquid chromatography, paper chromatography and thin layer chromatography. *See also* **High performance liquid chromatography**.

Liquid ion exchangers. Ion exchangers in the form of water-immiscible liquids possessing ionogenic compounds,[294] used to achieve selective exchangeability in solutions. Long-chain aliphatic amines and alkyl phosphonic acids are most commonly employed for this purpose. Heptadecyl phosphoric acid, $C_{17}H_{35}OPO(OH)_2$, is used in this way in solvent extraction systems, the protons of its hydroxyl groups being exchangeable for metal ions, and it can be regenerated by treatment with acids to release the metals. Such exchangers have been employed for extractions for most metals, including actinoids, lanthanoids and transition metals. Metals possessing several oxidation states can frequently be extracted more readily in one form than in another. Vanadium, for example, as vanadium(III) and vanadium(V) can be extracted from sulphate solution by primary amino liquid ion exchangers, but the vanadium(IV) form is only extracted by dialkyl phosphoric acids. Although they have been used for many years the full potential of liquid ion exchangers does not appear to have been realized.

Liquid–liquid chromatography. Partition chromatography systems employing both a liquid mobile phase and a liquid stationary phase. *See also* **High performance liquid chromatography** and **Paper chromatography**.

Liquid phase. The use of this expression can frequently be rather ambiguous and should be avoided if possible. In GLC it refers to the low-volatility liquid held as a thin film on the solid **support** where it is in direct contant with the gas (mobile) phase from which solutes separate according to their differing affinities. In LC the expression is used to mean the mobile solvent phase.

Liquid–solid chromatography. All chromatographic systems involving passage of a mobile liquid phase over a solid adsorbent. These are all classified as adsorption chromatography, and include conventional column methods, **dry-column chromatography**, and **thin layer chromatography**.

Logarithmic dilution method. Calibration of GC detectors for accurate quantitative analysis is best carried out by means of a special dilution sampling device for producing accurate concentrations of solute in the gas suitable for determining the detector response. In the device used for this purpose,[295] argon is passed through a nonvolatile liquid and a small amount of solute is injected into the headspace above the liquid. The solute partitions between the liquid and the argon gas, so that exit gas from the dilution vessel contains some solute when passed to the column and eventually to the detector. The amount of solute in the gas stream at any time can be calculated from the gas flow rate and the distribution constant, and the peak area recorded is plotted against the concentration.

The procedure is referred to[296] as the logarithmic dilution method as its equation shows that if the detector gives a linear response then a plot of log peak area against time will give a straight line. Its success is dependent upon a sufficiently large solute sample being placed into the headspace and a constant volume of exit gas being employed to achieve the progressive series of dilutions.

Lower limit of detectability (E_u). The smallest amount of a substance that can be detected with any certainty by a detector.[297] To be strictly accurate the substances employed for the determination should be specified, as this limit varies from one group of compounds to another. The values are established by measuring the signal intensities for progressively smaller and smaller quantities, then plotting a graph of signal against quantity, extrapolating down to a signal corresponding to double the detector noise level and reading the quantity corresponding to it. The resulting value is often expressed in parts per million (by volume).

Different detectors can be compared with each other[298] on the basis of the lower limit of detection, although such comparisons are really only strictly valid between detectors operating according to similar principles. *See also* **Detectors**.

Low voltage zone electrophoresis. Separation of high molecular weight substances by zone **electrophoresis** is best carried out under low voltage conditions (100–400 V). The method is unsuitable for low molecular weight compounds due to the diffusion which occurs during prolonged periods of separation. *See also* **High voltage zone electrophoresis**.

M

M. See **Migration value.**

Macroporous resins (Macroreticular resins). Special cross-linked **ion exchange resins** have been manufactured possessing not only the normal micropores but also macropores several hundreds of Angstrom units wide and large when compared with atomic distances. These macroporous resins are, therefore, highly porous with large internal surface areas easily accessible even to very large molecules. The available capacity of such resins is greater and exchange rates are better than for corresponding **microreticular resins.** *See also* **Pellicular coating.**

Mark-Houwink equation. Determination of average molecular weights by **gel chromatography** is dependent upon the use of substances with related molecular structures and known molecular weights for calibration. Hendrickson and Moore[299,300] devised a set of rules based upon 130 compounds in tetrahydrofuran solution which could be correlated into a single calibration line based upon chain length.[301] Other studies, however, have shown that this only applies for systems for which the value for *a* calculated from the Mark-Houwink equation, below, are nearly identical:

$$[\eta] = KM_r^a$$

The equation relates the limiting viscosity of the solution to the molecular weight of the solute.

A more exact approach is, however, obtained by employing the hydrodynamic size in solution, as in the **Flory–Fox relationship.**

Mass detector (Brunel mass detector). A detector[302] in which eluates from GC are directly weighed on a microbalance. It comprises a small cylindrical chamber, containing active charcoal, attached to the microbalance and on to which the eluate is adsorbed. The response is a function of the weight of the material adsorbed by the charcoal, the device acting as an **integral detector**. It has been shown to be[303] suitable for quantitative analysis, giving complete adsorption over wide ranges of conditions with a sensitivity comparable to the **katharometer**.

Mass distribution ratio (k). A term recommended[304] in place of the expressions capacity factor and partition ratio, and defined by

$$k = \frac{\text{solute concentration in the stationary phase}}{\text{solute concentration in the mobile phase}}$$

For GC, in which allowance has to be made for the pressure gradient correction factor,

$$k = K_D \frac{V_S}{jV_I}$$

For HPLC the more simple relationship is

$$k = K_D \frac{V_S}{V_I}$$

Obviously the value for k can be increased by increasing the volume of the stationary phase in the column relative to the mobile phase.

The relationship between k and other parameters is given by

$$k = \frac{t'_R}{t_m} = \frac{V'_R}{V_I}$$

Mass transfer term (C). The third term of the **van Deemter equation**[305], usually given in the following form:

$$C\bar{u} = \frac{8}{\pi^2} \frac{k}{(1+k)^2} \frac{d_f^2}{D_L} \bar{u}$$

Its importance rests in the fact that it includes the stationary phase thickness (d_f) as a squared term, and therefore becomes the major factor in determining the **height equivalent to a theoretical plate** at high gas velocities. It is of similar importance in the calculations for theoretical plates for HPLC, zone spreading being reduced by employing thin stationary phase films on the supports and thus increasing the rate of desorption.

Matrix. In chromatography this word has acquired two main usages. It is frequently employed with reference to the column packing material, i.e. the support plus stationary phase in partition chromatography or the solid adsorbent in adsorption chromatography, in the sense that 'substances penetrate the solid matrix'. Its other use is in connection with the combination of column packing (ion exchange resin or gel) plus the solvent within the interstices of this packing; in this latter case the gel (or resin) and solvent are considered to form a continuous matrix.

Mayer–Tompkins theory. A theory for ion exchange systems which predates that due to Glueckauf (*see* **Glueckauf equation**), which in turn served as the forerunner of the **van Deemter equation**. The theory[306] was a development of the stepwise distribution system for theoretical plates in chromatographic columns first introduced by Martin and Synge.[307] The theory developed by Mayer and Tompkins enabled predictions to be made of the number of plates necessary to achieve a particular level of purity in the separated solutes. Their theory, however, was only suitable for systems in which distribution coefficients of solutes between mobile and stationary phases remain constant throughout the column, and for high efficiency columns.

McReynolds' constants. A set of tables obtained as differences (ΔI) between the **Kováts indices** for a wide range of substances on various liquid phases. The constants were compiled by McReynolds[308] in order to compare the separating ability of stationary phases in GLC. The larger the McReynolds' constants the greater the retention time on the particular phase specified. The constants provide information about the relative efficacies of liquid phases and enable comparable phases to be selected.[309]

Mesh sizes. Much confusion exists over particle sizes employed for column packings as these are sometimes quoted in mesh sizes and sometimes in micrometres. Table 5 juxtaposes the two sets of figures, and the relationship can be seen more clearly in Figure 35.

Table 5

British and US standard mesh sizes	Diameter (μm)
10	2000
20	840
30	590
40	420
50	297
60	250
70	210
80	177
90	164
100	149
120	125
140	105
160	92
180	82
200	74
400	37

Methyl silicone gums. Some of the best general purpose liquid phases for partition chromatography. Because of their low volatility they can be employed for operating temperatures up to 250 °C at a 10% loading, and are suitable for separations of both moderately polar and strong polar compounds. These gums are members of the extensive series of organo-silicon compounds possessing a general three-dimensional structure based upon the following type of silicon–oxygen network:

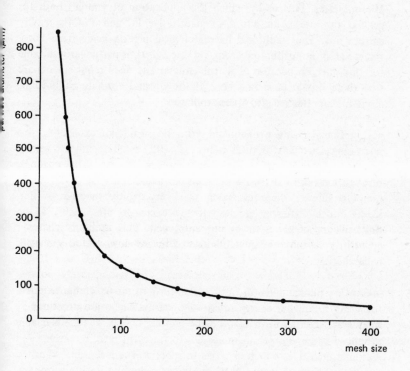

Figure 35. Relationship between mesh sizes and particle diameters

Microcapillaries. Precision-made capillary tubes about 3 cm long which will retain a predetermined volume of solution by capillary action after being dipped into the sample solution. They are one of the means used in quantitative TLC for the application of small quantities of sample which have been accurately measured. When the tip of the capillary is placed on the TLC plate, the solution is drawn from the tube by the adsorption of the plate surface. These capillaries are available in various sizes with capacities from 1 to 10 μL. *See also* **Micropipettes.**

Microcoulometric detector. *See* **Coulometric detectors.**

Micropipettes. Devices by which the application of samples both for general purpose work and for quantitative PC and TLC is readily carried out. The traditional haematological pipette made from borosilicate glass and calibrated from 5 μL to 25 μL is particularly suitable for this purpose because of its robustness. The main requirements are that there should be a fine bore at the tip and accurate calibrations along the pipette. *See also* **Microcapillaries.**

Microreticular resins. Cross-linked synthetic organic **ion exchange resins** possessing narrow structural networks with small openings approximately corresponding to ordinary molecular sizes. The channels through the resin matrix are between 0.5 and 5 nm wide (5–50 Å). In nonpolar solvents these resins only swell very slightly; their narrow networks impair diffusion and lead to low exchange rates, such that the full capacity of the resin is not employed. This problem has been greatly reduced by the development of **macroporous** resins and **pellicular coatings.**

Microwave plasma detector. A modified form of the **photometric detector** using the higher energy microwave plasma excitation produced by microwave excitation of argon or helium carrier gas.[310] Free electrons produced cause the solute molecules to fragment into free atoms and diatomic molecules which give rise to spectra. The technique is limited due to interference from CN, C_2 and CH species, but is suitable for the selective study of compounds containing halogens, phosphorus and sulphur atoms, the limits of detection being about 1×10^{-10} g.

Migration value (*M*). In zone **electrophoresis**, the distance travelled in cm from the origin during a period of 1 s at a field strength of 1 V cm^{-1}. Values obtained are very small, for example a substance moving 8 cm in 12 h when 50 V are applied across a 30 cm strip has a migration value

$$M = \frac{8}{12 \times 3600} \times \frac{30}{50} \approx 1 \times 10^{-5} \text{ cm}^2 \text{ s}^{-1} \text{ V}^{-1}$$

Mixed-bed column. Total removal of an ionic compound by ion exchange methods requires the use of both **anion** and **cation exchangers**. In a mixed-bed column the two resins are mixed together in the proportions necessary to remove both positive and negative ions simultaneously. This means that, if the metal ion is being replaced by H$^+$ and the anion by OH$^-$ from the resins, the solution does not acquire the strongly acidic or basic character that initially occurs if separate anion and cation resins are employed.[311,312]

The most common application of a mixed-bed column is in water softening, the inorganic impurities in the water being replaced by the H$^+$ and OH$^-$ ions from the column.

Such columns present a problem when it is necessary to carry out regeneration. Initially the two resins have to be separated by a backflow of water and then separately regenerated by the introduction of appropriate solutions at the two levels corresponding to the respective heights of the separated resins. After regeneration the resins are remixed by blowing compressed air upwards through the bed.[313]

Mixed solvents. Development of chromatograms is frequently achieved by using solvent mixtures consisting of a polar solvent and a nonpolar solvent, for example 1–5% ethanol in cyclohexane. Such mixtures are frequently established by trial and error for one particular chromatographic separation. The use of a small proportion of polar solvent and a high proportion of nonpolar solvent is most common in paper and thin layer chromatography, as a high proportion of polar solvent frequently leads to all compounds moving with the solvent front. In some instances these solvent systems for TLC may consist of three or more mixed solvents in different proportions. However, the advantages of such systems seem doubtful as the solvent may undergo partial separation as it penetrates the adsorbent.[314,315]

Mixed solvents are also employed in column chromatographic procedures, but in these instances it is more common to make use of **gradient elution** procedures based upon an **eluotropic series** of solvents.

Mixotropic series. *See* **Eluotropic series.**

Mobile phase. That phase in any chromatographic system which carries the solutes along and over the column material or surface on which the separation is being carried out. For **gas chromatography** the mobile phase is commonly an inert gas such as helium, argon or nitrogen, while in **liquid chromatography** it may be one of the many solvents or mixtures of solvents depending upon the form of the separation. *See also* **Stationary phase.**

Molecular diffusion term. The second term in the common representation of the **van Deemter equation.**[316] The molecular diffusion contribution to the **height equivalent to a theoretical plate** is equal to

$$\frac{\gamma D_M}{\bar{u}}$$

where γ is known as the labyrinth factor, and for GLC has a value of 0.5–1.0, accounting for the irregularities in the mobile phase flow through the column particles. At low flow rates the height equivalent to a theoretical plate is mainly determined by this term, while at high flow rates its contribution becomes virtually negligible. This term was called the axial molecular diffusion term in the original presentation of the van Deemter equation.

Molecular radius of gyration. *See* **Stokes' radius.**

Molecular sieve. A term first introduced by McBain[317] after his observation that, while the **zeolite** Chabazite adsorbed alcohols, other substances such as benzene and acetone were excluded. The term is now widely used to refer to fine porous solid substances that can be used to carry out separations based upon differences in molecular dimensions, particularly synthetic alkali metal alumino silicates.

Molecular sieves are particularly employed for drying gases, such as those in refineries, to < 1 p.p.m. of moisture using zeolites with 3–4 Å pore size. The process is assisted by the presence of cations in the crystal structure of the zeolite which attract polar molecules and possess a high capacity even at low partial pressures.

Zeolites have also been used to remove water from alcohols, ethers and amines. The desulphurization of petroleum is achieved by the use of molecular sieves with a 13 Å pore size, while those with a pore size of 10 Å are used to remove water and hydrocarbons from carrier gases before entering the columns in gas chromatographs.

Breck[318] has differentiated between two forms of molecular sieving. Total adsorption or rejection occurs if one species diffuses fully into the molecular sieve while a second species is totally unable to diffuse into the structure. Partial selectivity for the components of a binary mixture occurs if the two species diffuse into the molecular sieve at different rates. Both methods of separation are frequently used.

Molecular weight estimation. Gel chromatography can be used as a method for estimating molecule weights for high molecular weight materials such as proteins.[319] This is made possible by the substantially linear relationship between the logarithm of the molecular weight and the various solute migration indices. In most cases the elution volume itself is suitable for this purpose, the linear relationship extending over a wide range of molecular weight values.[320] Separate calibration curves are needed for different molecular species and different gel permeation media, as the molecular weight–elution relationship differs depending upon the general features of the molecule. The molecular weight of an unknown substance can be established by ascertaining where its elution volume falls on the appropriate calibration curve. *See also* **Selectivity curve**.

Moving bed process. Although the moving bed process is normally classified as a chromatographic process this is sometimes disputed as both the phases are mobile. It is, however, best considered as a form of continuous column chromatography which involves moving the chromatographic bed (the normal stationary phase) in the opposite direction to the mobile phase such that, by the choice of appropriate flow rates, one component or group of components moves with the mobile phase relative to the other components which move in the direction of the moving bed.

In the apparatus described[321] for this a 2.5 cm internal diameter column was used and the gas flow was from the lower end upwards, while the moving bed was added continuously at the top of the column. This process has been sucessfully used for the separation of large quantities of binary mixtures.[322]

Moving boundary electrophoresis. A technique which does not usually result in a good separation of ions and is mainly of historical importance, the earliest electrophoretic studies being of this type.[323] Essentially it involves the application of a current to the end buffer compartments of a U-tube between which is contained a solution with two ions requiring separation. When the current is applied the faster moving ion migrates clear of the slower moving ion. Although clear zones of each ion are formed there is also a large area of overlap.[324]

Moving wire detector. *See* **Wire transport detector.**

Multiple development. A procedure used in PC and TLC, devised as an alternative to the column technique of **gradient elution** which is not possible with solvents of increasing polarity. In this approach the chromatogram is developed along part of its length with one solvent, then removed from the tank, dried and redeveloped for a greater extent with another solvent.[325] The exercise can be repeated several times with solvents of increasing polarity if required, although usually only two solvents are used for any particular chromatogram. The technique is often employed if it is found that the first solvent only causes the compounds in the sample to move very slowly with little separation on the chromatogram.

Multiple layer chromatography. *See* **Dual layer chromatography.**

N

N,0-bis-(trimethylsilyl)-acetamides. *See* **Silylation.**

Net retention volume (V_N) (Absolute retention volume). The value for the retention volume in GC corrected for the difference in pressure between the head of the column and the outlet. It is given by the relationship

$$V_N = jV_R'$$

where V_R' is the adjusted retention volume and the value for j is obtained from the inlet and outlet pressure correction equation:

$$j = \frac{3}{2}\left[\frac{\left(\dfrac{p_i}{p_o}\right)^2 - 1}{\left(\dfrac{p_i}{p_o}\right)^3 - 1}\right]$$

Nitrogen detector. *See* **Thermionic detector.**

Noise. Background signal fluctuations arising from a detector[326] and the associated circuits and recorded in the baseline of a chromatogram. Such background variations cannot be totally eliminated without reducing the signal intensity and affecting the detector sensitivity. Changes in the level of the background noise can frequently be a useful guide to deterioration in the detector. For a detector signal to be recognized as valid it has to be twice the noise level. *See also* **Signal-to-noise ratio.**

Nominal linear flow (F). The **flow rate** of the mobile phase divided by the cross-sectional area of the column, producing a value expressed in cm min^{-1}.

Normal distribution. *See* **Gaussian distribution.**

O

Open tube columns. Capillary columns used in GC and characterized by having an unobstructed axial gas flow region; mixtures are separated through interaction with materials on the walls of the tube. These coatings are classified as three types: porous layer open tube columns (**PLOT columns**); support coated open tube columns (**SCOT columns**); and wall coated open tube columns (**WCOT columns**).

Optically active resins. Separation of optically active isomers has been carried out[327,328] by the incorporation in an ion exchange resin of fixed optically active groups which can interact with the isomers in the mixture passing down the column.

Optically active **cation exchangers** have been made by condensing phenyl succinic acids with formaldehyde and other optically active acids.[329] Although enrichment of optical isomers has been achieved by such resins they have not been overwhelmingly successful for the purpose for which they were designed.

Orange dextran. Determination of the void volume (*see* **Interstitial volume**) for gel chromatography is most commonly carried out with **Blue dextran**, but for thin layer gel chromatography this is not suitable as the small amount of blue colour is almost invisible on the greyish gel. Of the various dextran-dye combinations[330] prepared, Miller[331] found Orange dextran to be particularly suitable for thin layer gel permeation chromatography. It is made by dyeing dextran type 2000 with Pocion Brilliant Orange 2RS using a 1:1 dextran:dye ratio. The product absorbs at 494 nm and as little as 1 μL of a 0.2% w/v solution can be seen on the chromatogram.

Outer volume (V_o). *See* **Interstitial volume**.

Over-run chromatogram. A paper or thin layer chromatogram which is allowed to run for an extended period so that the solvent front reaches the end of the surface. Such a chromatogram is unsuitable for the calculation of R_F values. It is not uncommon to deliberately create an over-run chromatogram in descending PC in order to increase separation on slow moving closely spaced compounds.

P

Packing. The adsorbent, gel or solid supports (plus any deposited stationary phase) used in a column chromatography procedure. In most instances packings consist of particles of a limited size range, for example between 80 and 100 or 100 and 120 **mesh size**, and sometimes much finer than this. Even when coated with a stationary phase for partition chromatography they should be dry to touch and free flowing in order that they can be poured into the column under suction. The packing material should be continuous through the column with no cracks or breaks that will lead to diffusion and deterioration of the separation.

Paper. A wide variety of papers is available for chromatography, the great majority being very pure cellulose papers. Trace impurities of carbohydrates, fatty acids and inorganic substances do exist but rarely interfere seriously with chromatographic separations. The papers are designed to give fairly rapid flow rates with a large solvent up-take and high mechanical strength. Chemically modified papers are also produced suitable for ion exchange and adsorption chromatography.

Paper chromatography. Although the introduction of the modern form of paper chromatography is justifiably credited to Consden, Gordon and Martin,[332,333] the first real paper chromatograms were produced by Runge[334,335] in the nineteenth century, and photographs of his circular chromatograms have been reproduced in more recent publications.[336,337] Consden and his colleagues were also aware of the earlier work of Schönbein[338] and of Goppelsröder[339] involving adsorption on strips of cellulose. Paper chromatography is essentially a partition process (although adsorption and ion exchange also have small roles) in which the paper acts both as the support and, by virtue of the water held in its fibres, as the stationary phase. Alternatively the paper may be impregnated with some other suitable partitioning solvent.

Chromatograms are run either as **descending chromatograms**, in which the upper edge is in a trough of solvent, or as **ascending chromatograms** with the lower edge dipping in the solvent. Paper chromatography is most suitable for separations on the mg-μg scale, although it is now more common to use **thin layer chromatography** wherever possible as it is more rapid.[340]

Partition. If a substance X, soluble in two immiscible solvents A and B, is shaken with the solvents in a separating funnel, it will partition itself between the two solvents in a constant ratio irrespective of the quantity of solute added (as long as neither solvent is saturated). That is:

$$\frac{\text{concentration of X in phase A}}{\text{concentration of X in phase B}} = \text{constant}$$

This process of partition between two solvents is the basis of **counter-current distribution** and of all forms of **partition chromatography**. *See also* **Distribution constant**.

Partition chromatography. Lack of success in bringing about separations by traditional partitioning methods led Martin and Synge[341] to attempt partitioning on columns in which one of the liquid phases was held stationary on a granular solid support while the other was allowed to flow freely, under gravity, down the column. Their work was the start of what is now known as partition chromatography in which separation of a mixture is achieved by passage of a mobile phase (liquid or gas) over a stationary liquid phase (which may be on paper, solid particles or the walls of a capillary tube). Liquid–liquid, gas–liquid and paper chromatography are all classified as partition processes. *See also* **Adsorption chromatography**; **Gel chromatography**; **Ion exchange chromatography**.

Partition coefficient. *See* **Distribution constant**.

Partition ratio. *See* **Mass distribution ratio**.

PC. *See* **Paper chromatography**.

PEGs. *See* **Polyethylene glycols.**

Pellicular coating. A thin film or pellicle of ion exchange resin coated on a solid inert centre core. The idea was introduced by Horvath *et al*[342] to achieve rapid exchange rates on ion exchange resins for HPLC. They used a thin film of styrene-divinylbenzene and some benzoyl peroxide polymerized onto glass beads at 90 °C, and further substituted the resulting polymer to give the normal type of **anion** or **cation exchanger.** This produced the solid core surrounded by the film of cross-linked ion exchange material stable to organic solvents, high buffer strengths and temperatures up to 80 °C. Such resins have low exchange capacities, 5–10 microequivalents g^{-1}, but high efficiencies and rapid exchange rates. The procedure has since been widely applied to give many different forms of **porous layer beads**, mainly using inert siliceous particles of controlled surface porosity as the support. These range in size from about 50 nm to 200 nm diameter, the cross-linked polymer being formed in the cavities of the support bead.[343] The surface areas of resulting stationary phases range from 8 to 15 m^2 g^{-1}.

Peristaltic pump. *See* **Pump (peristaltic).**

Permeation chromatography. Chromatographic separations which are based upon exclusion effects such as the size or shape of molecules (molecular sieving) or on charge (ion exclusion). The most widely used permeation technique is that of **gel chromatography** in which the stationary phase is either a swollen gel or a rigid three-dimensional silica or polymeric matrix.

Permselectivity. Permeation of certain ionic species in preference to other ionic species through ion exchange membranes.

Phase ratio (β). The ratio of the volume of the mobile phase to the volume of the stationary phase in partition chromatography:

$$\beta = \frac{V_M}{V_s}$$

Phase transformation detector. *See* **Wire transport detector.**

Phosphorus detector. *See* **Photometric detector** and **Thermionic detector.**

Photodensitometer. *See* **Densitometer.**

Photoionization detector (PID). A detector in which photons of suitable energy are used to cause the complete ionization of the solutes in an inert carrier gas.[344,345] The most common source of the ionization is ultraviolet irradiation[346] in a detector cell containing a hollow cathode through which the carrier gas and solute enter.

In a more recent design,[347] separate discharge and detection chambers are used, with a lithium fluoride window between them which allows ultraviolet radiation of 105-110 nm to pass. Ionization of the solute produces an increase in current from the detector which is amplified and passed to the recorder. Another development of this type of detector[348] has been in a portable gas chromatograph for the purpose of trace gas analysis in polluted and industrial environments. Detection limits at the parts per billion level are possible for some vapours.

Photometric detector (Flame photometric detector; Flame emission detector). A detector[349] which is based upon the **flame ionization detector**, but uses a photocell to monitor the light obtained from burning the eluate from a GC column in a hydrogen-rich flame. It can therefore be made specific for sub-microgram amounts of halogens,[350] phosphorus and sulphur.[351] The selectivity for phosphorus is obtained by using a filter to detect the HPO emission at 526 nm, and for sulphur by a similar filter for the S_2 emission at 394 nm. Sensitivities for these elements are about 1×10^{-12} g s^{-1}.

The detector gives a constant response for members of homologous series, but not between different series of compounds. *See also* **Selective detectors.**

PID. *See* **Photoionization detector.**

Piezoelectric detector. Various forms of the piezoelectric detector have been devised and this is currently one of the most promising fields of detector research. In these devices[352,353] a small crystal, such as quartz, is coated with a stationary phase material and placed at the end of the GC column. As solutes are eluted from the column they pass over the coated crystal and give rise to sorption and desorption on the crystal with corresponding frequency changes in the piezoelectric effect related to the quantity of material sorbed.

Quartz flakes with coatings of **squalane** or **polyethylene glycols** and vibrating frequencies up to 9000 MHz have been found to be particularly good,[354] and detection limits as low as 1×10^{-7} have been obtained.

Planimeter. A device that has been used for many years to measure areas of irregular shapes. In chromatography it has been applied to measure the areas of chart peaks for quantitative studies on chromatograms. The planimeter consists of two arms with a digital meter between them. The end of one arm is held in a fixed position while the end of the other arm is traced around the area (Figure 36). Much of the success and accuracy of the method depends upon the skill of the operator. It is, however, a rather tedious method of measurement and little more accurate than **triangulation**.

Plasma. An approximately electrically neutral ionized gas containing simultaneously a variety of positive and negative ions, electrons and neutral molecular and atomic species[355] created as a result of ionization of compounds. This ionization can occur in a flame, naturally in the ionosphere (aurora borealis) or in electric lights by discharge through gases.[356] For the plasma to be maintained, a high a.c. or d.c. electric field must be applied to prevent recombination of the charged species. Several forms of GC detector (e.g. **flame ionization** and **radio-frequency detectors**) are based upon the creation and maintenance of a gas plasma.

Plate height. *See* **Height equivalent to a theoretical plate** and **Theoretical plate**.

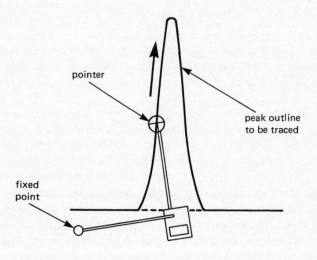

Figure 36. Planimeter.

PLOT columns. Since **capillary columns** were first introduced for GLC various modifications have been made in order to increase the capacities and the efficiencies. The porous layer open tubular (PLOT) columns were developed[357] in order to extend the application of capillary columns to GSC, and consist of capillary columns with very fine layers of adsorbent, such as alumina or charcoal, deposited on the inside wall,[358] but maintaining a free path along the axis of the tube.[359,360] Such a column has a height equivalent to a theoretical plate of 1-6 cm at optimum flow rates of 40-70 cm s^{-1}.

Poisson distribution. Although most chromatographic curves correspond to the **Gaussian distribution**, nonideal conditions can lead to peaks which correspond instead to the Poisson distribution given by the following equation:

$$P(x) = e^{-\mu} \frac{\mu^x}{r!}$$

in which the mean and the variance both equal μ.

If μ is large the curve approaches that for the Gaussian distribution. The binomial distribution of $(q + p)^n$ approaches the Poisson distribution when $n \to \infty$ and $p \to 0$.

Polarity. A term which is employed in a very loose sense in chromatography to refer to the ability of solids to serve as adsorbents and to refer to the affinity that solutes have for the stationary phase, as well as to indicate the ability of a solvent to maintain a solute in solution.

A polar stationary phase is one that has a strong affinity for solutes. In adsorption chromatography Grade I **alumina**, for example, is considered to be a very polar adsorbent as compounds only move very slowly along columns packed with it. The effects of a polar column are best overcome by the use of a polar solvent such as methanol or ethanol, which will move the compounds along the column or TLC plate even when strongly held by a polar adsorbent. In the same sense a polar compound is one which will have a strong affinity for a polar stationary phase.

It should be emphasized that while polarity in the chromatographic sense is associated with functional groups on surfaces, in solvents and on compounds it is very empirical and not necessarily directly related to the more quantitative concepts of physical chemistry involving calculations of dipole moments. Attempts have been made to reduce this empirical aspect in the case of solvents by the use of **solubility parameters** to construct suitable **eluotropic series**.

Polarographic detectors. Detectors which are based upon the measurement of currents between a polarizable and a nonpolarizable electrode and have been specially developed for use with HPLC. These detectors operate at constant voltage, the current flowing being recorded against time. Both dropping mercury[361,362] and carbon-silicone rubber[363] have been used as the polarizable electrode. In the latter case the electrode consists of a wire dipped into a pool of mercury held in a glass tube by a silicone rubber disc impregnated with carbon (sometimes called the carbon membrane electrode). In the detector the solvent flowing from the HPLC column passes into a small space below the silicone rubber disc, and the current between this and a platinum electrode is measured. Using a cell volume of 10 μL a sensitivity of 10^{-9} g cm^{-3} has been claimed.

Polyacrylamide gels. Neutral hydrophilic polymers obtained by the co-polymerization of acrylamide with N,N'-methylene-bis-acrylamide.[364] They are sold under the trade name Bio-Gel P (manufactured by Bio-Rad Laboratories). Ten grades of Bio-Gel P are available with **exclusion limits** ranging from about 1800 to 400 000. Their great value lies in their insolubility in organic solvents.

Polyethylene glycol adipate. The most widely employed of the many terminally substituted **polyethylene glycols** that have been tried as stationary phases for GLC. It is used up to a 10% loading on the support, has an operating limit of about 200 °C, and has been found to be particularly suitable for separating unsaturated aliphatic compounds.

Polyethylene glycols (PEGs). Materials, commonly referred to as polyethylene glycols, corresponding to the general formula

$$HOCH_2CH_2-[O-CH_2CH_2]_n-OCH_2CH_2OH$$

They are available with molecular weights covering the range from 106 to 6000 ($n = 120$). Their properties change from high boiling liquids for the short chain compounds to hard waxes for the long chain compounds.

PEGs with molecular weights 400–1500 are most commonly used as stationary phases for GLC[365] at a 10% loading with an operating limit of about 130 °C, suitable for separations of polar substances such as amines, alcohols and esters. These materials are also sold under the trade name Carbowax.

Polystyrene gels. Gels formed from polystyrene cross-linked with 2% divinylbenzene. Very good separation of lipids[366] have been achieved by gel chromatography employing such gels. They are very rigid, their porosity depending upon the dilution of the original polymerization mixture. Gels with exclusion limits as high as 8×10^6 have been prepared[367] by this method.

Porapak. The trade name for a wide range of cross-linked polymer beads formed from combinations of divinyl benzene and styrene or ethyl vinyl benzene and suitable for use as adsorbents in GSC. They hae surface areas within the range 100–600 m^2 g^{-1} and can be used at temperatures up to 250 °C without structural deterioration taking place. They have been shown to be of particular value in separating water from organic substances.

Porous layer beads (Controlled surface porosity supports). To reduce the effect of the **mass transfer term** in the **HETP** values in HPLC it was found necessary to devise new forms of solid supports for stationary phases.[368] The more general purpose porous layer beads were developed from the **pellicular coatings** and consist of a small solid core covered by a uniform thin layer of sorbent. This outer layer is either an adsorbent deposited on the surface of a solid bead or, in some cases, an etching on the surface of the bead itself such that the porosity is carefully controlled. In all cases the thin surface layer serves to reduce the migration distance for the solute molecules. As a result high separation velocities are possible while still maintaining conditions near to equilibrium between the column phases. Such column packings have the advantage of retaining their spherical shape under the applied pressures within the HPLC columns.

Porous layer open tubular columns. *See* **PLOT columns.**

Porous silica gels. *See* **Silica aerogels.**

Potential difference radio-frequency detector. A detector which measures changes in potential difference between a probe, inserted into a radio-frequency plasma produced with helium carrier gas, and a capacitor plate. The changes in the potential occur whenever the composition of the gas changes as solutes are eluted from the column.[369] Although it is similar to the **frequency difference radio-frequency detector** it is claimed that its results are not as reproducible. The limit of detection is < 10 p.p.m. with a sensitivity of 1×10^{-8} g s^{-1} for carbon monoxide, carbon dioxide, nitrogen and sulphur dioxide.

Preparative chromatography. Although most chromatographic separations are essentially batch processes, many attempts have been made to extend chromatography to large-scale preparative separations.[370] These large-scale methods fall into two main categories: either separation of large quantities of sample by repetitive injection of many small volumes, usually by automatic sampling and collecting devices, or truly continuous processes employing modified columns. **Helical flow columns** come into the first category, as separate injections are made into a series of columns; the **moving bed process** and **radial flow columns** are examples of the second approach.

So far most of the work on preparative chromatography has been in GC, but prepartive HPLC is now being developed using large diameter columns with the **porous layer beads** originally introduced for normal HPLC columns.[371] Thick layer techniques are used to obtain larger quantities of separated materials from **preparative thin layer chromatography**.

Preparative electrophoresis. *See* **Continuous electrophoresis.**

Preparative thin layer chromatography. Although conventional TLC is used with samples containing up to 10–20 μg of sample, the separation of larger quantities of material can be carried out by what is frequently called thick layer chromatography. This method uses specially prepared plates of adsorbent up to 5 mm thick on a full size 20 cm plate. Such plates have to be specially dried to avoid cracking and flaking. By means of a special syringe mounted in a carriage the sample solution is applied in a continuous line about 1 cm from one end of the plate. Repeated applications can be made until the required amount has been deposited on the starting line. The plate is then developed in the normal TLC manner, and when separation is complete the compounds can be scraped from the plate and extracted from the adsorbent. Up to 1 g of material can be separated in this manner, but for larger quantities **dry-column chromatography** is preferable.

Pressure gradient. The difference in pressure between the inlet and outlet of GC and HPLC columns. Although the ideal state with no pressure drop would give an inlet-to-outlet pressure ratio of $P_i/P_o = 1$, this is never achieved even in **WCOT columns**. In the mobile phase carrier gas of GC, the difference in pressure between the inlet and outlet means that the **net retention volume** can only be calculated if the **adjusted retention volume** is multiplied by the column pressure gradient correction factor, j:

$$j = \frac{3}{2} \left[\frac{\left(\frac{p_i}{p_o}\right)^2 - 1}{\left(\frac{p_i}{p_o}\right)^3 - 1} \right]$$

This particular problem does not arise to the same extent in HPLC as any volume change is very much less due to the low compressibility of liquids. However, in this case the lower pressure at the outlet end of the HPLC column can lead to dissolved gases being released from the solvent and producing misleading detector readings. In this case the problem is dealt with by using **back-pressure devices**.

Pressure programming. *See* **Flow programming.**

Programmed temperature gas chromatography. *See* **Temperature programming.**

Pulse dampers. *See* **Dampening devices.**

Pulseless pumps. *See* **Pump (piston)** and **Pump (pneumatic)**.

Pump (peristaltic). In its simplest form, a series of rollers projecting from the circumference of a rotating disc and resting on a flexible tube carrying the solution (Figure 37). Rotation of the disc causes the rollers to press in turn on the tube, forcing solvent along by a massaging action in one direction. Such pumps are used in chromatography for the continuous constant addition of solvent to conventional liquid columns, and particularly for maintaining the flow of buffers in some forms of **amino acid analyser**. They have the great advantage that the flow rate is widely and accurately adjustable to deliver from a few cubic centimetres to several litres per hour. In addition they move liquids along tubes without the liquid coming into contact with any of the parts of the pump. This has been of particular importance in dealing with biological solutions to prevent contamination by lubricants and to maintain sterility.

Pump (piston). A device which is really like a giant-sized hypodermic syringe designed to force solvent from a reservoir along the capillary tube to the HPLC column. It is a pulseless pump in which the solvent is held in a fixed volume (500 cm^3–1 L capacity) thick walled stainless steel reservoir. By means of a piston, which is driven down into the reservoir at a programmed rate, the solvent is forced up through a capillary hole in the axis of the piston and along a tube to the column. By the operation of two reservoirs together with a mixing chamber this system can be readily used for **gradient elution**. The main disadvantages are the limited capacity of the reservoir and the time taken to change from one solvent system to another.[372]

Pump (pneumatic). A pump used for HPLC in which no moving parts come into contact with the solvent as the pressure is exerted by means of a high-pressure supply of inert gas.[373] In many ways this can be considered as being similar to the piston pump with a gas as the piston in place of a metal rod.[374] Such pumps are also pulse free, and have the added advantage that the fixed-volume reservoir can be immediately refilled from an external reservoir by means of a pressure release valve.

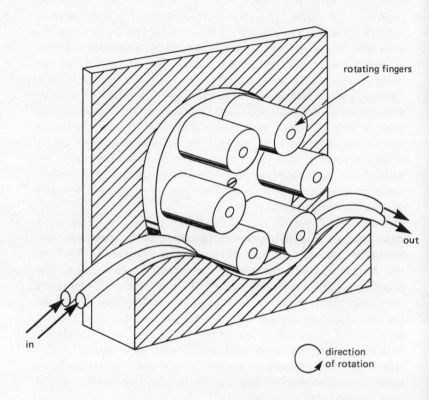

Figure 37. Peristaltic pump

Pump (reciprocating). A pump which incorporates a system of ball valves to prevent back-flow of solvent on the withdrawing stroke of the piston. There are definite disadvantages in the use of the traditional type of reciprocating pump for forcing solvents along the flow lines for HPLC.[375] Movement of the piston in one direction draws solvent from the reservoir into an intermediate flow chamber; on the forward stroke this solvent is then forced from the chamber into the flow line to the column.[376] As a result the solvent is delivered in a series of pulses which can cause fluctuations in some HPLC detectors. Although some of the pulsation is damped during the passage of the solvent through the flow lines, it is usually necessary to eliminate any undesirable background by means of some form of **dampening device** at the side of the flow line.

Pyrolysis gas chromatography. Nonvolatile materials that cannot normally be studied by GC procedures can be made to give chromatograms which serve as a means of identification by the process of thermal fragmentation known as pyrolysis.[377]

This consists of heating the material at a temperature sufficient to cause it to decompose into volatile low molecular weight products. These products of pyrolysis are transferred by some means to the column head and separated under defined column conditions. The chromatogram obtained serves as a 'fingerprint' for the material under the specified pyrolysis and chromatographic conditions. Adhesives, paints, plastics and rubbers have all been characterized by this method.[378,379] Its reliability for identification purposes is greatly dependent upon the reproducibility of the rate of pyrolysis of the sample; the best results are obtained by flash pyrolysis in which the maximum pyrolysis temperature is attained in a fraction of a second.[380]

The most common form of pyrolyser used in pyrolysis GC consists of a wire filament[381] that can be rapidly heated by high frequency induction heating to a fixed temperature (between 250 and 800 °C), within 0.02–0.03 s to give reproducible thermal decomposition of samples which have been deposited as thin coatings on the wire.[382] The pyrolyser is attached to the gas chromatograph at the injection port above the column head by removal of the septum and its retaining screw. As the sample is pyrolysed the products are swept straight into the column and separated. This form of pyrolysis unit (Figure 38) has the advantages that pyrolysis is rapid and only small quantities of material are required, while secondary decomposition of primary products is avoided. The limiting temperature

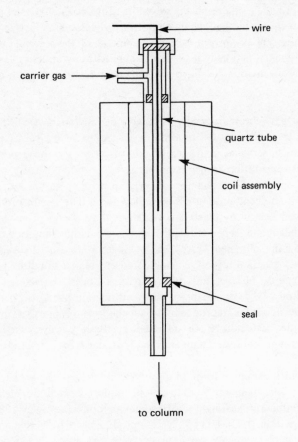

Figure 38. Curie point pyrolyser

(Curie temperature) is dependent upon the nature of the ferromagnetic wire used.[383]

Alternative procedures have included pyrolysis in a small boat in a chamber above the column head by employing external heaters,[384] and others in which the pyrolyser is separate from the chromatograph, the samples being taken from the atmosphere above the pyrolysed material and then injected onto the column. In some cases these systems suffer from the disadvantage that temperature increases are slow, and reproducible chromatograms that can serve as 'fingerprints' of the pyrolysed compounds are difficult to obtain.

R

Radial flow columns. A means of performing **continuous chromatography**, suitable for GC, the mixture being separated by a combination of the normal sorption/desorption processes and centrifugal forces.[385] The mixture is fed continuously to a point at the hollow centre of the stationary phase which is in the form of a short squashed column, disc shaped rather like a millstone. The stationary phase solid disc and a series of collectors around its circumference are rotated at a constant rate as the sample is fed in and the carrier gas forced through. As a result of the combination of forces acting upon the solutes, the least sorbed compounds follow the shortest curved paths to the circumference of the disc and the strongly sorbed compounds follow the longest paths, the separate substances being collected at fixed points on the rotating disc.

Radioactivity detector. Detection of radioactive substances separated by chromatographic means has always been possible by standard counting devices with TLC and with extruded columns. The procedure has been extended[386] as a specific method for detecting labelled compounds separated by HPLC. In this approach small scintillators are placed in contact with the column effluent, as radiation is emitted by the solutes these are detected by the scintillators and the light pulses from the scintillators are detected by a photomultiplier, amplified and recorded.

Radio-frequency detector. Any of several detectors[387] that respond to changes in either the dielectric constant of a radio-frequency **plasma** or the potential difference between capacitor plates arising from variations in the composition of the carrier gas as solutes are eluted from the GC column. Helium is the carrier gas most commonly employed[388] for this type of detector. Under baseline conditions a constant discharge is maintained giving a steady continuous voltage; passage of a solute causes a reduction in the discharge and a corresponding decrease in the voltage.

Radius of gyration. *See* **Stokes' radius.**

R_b **value.** A value representing in general terms the movement of one compound relative to a standard material on a paper or thin layer chromatogram:

$$R_b = \frac{\text{distance moved by sample spot}}{\text{distance moved by reference standard}}$$

For this determination the standard may be incorporated into the solution of the substance under analysis, or run as a separate spot alongside on the same chromatogram.

One example of this has been the use of Butter Yellow[389] (p-dimethylaminoazobenzene) as a reference standard on TLC plates for the comparison of 3,5-dinitrobenzoates obtained from reactions of alcohols and phenols with 3,5-dinitrobenzoyl chloride. The R_b value is obtained from the ratio

$$R_b = \frac{\text{migration distance of 3,5-dinitrobenzoate}}{\text{migration distance of Butter Yellow}}$$

Recycling separations. Although originally applied to GC,[390] recycling methods have been extended to HPLC, particularly using gel chromatography columns.[391] The purpose of the procedure is to gain the advantages of extended columns on a short column by recycling eluate fractions as they are eluted from the column. It involves linking the detector outlet to the column head with the mobile phase inlet. In HPLC this has to be done via the solvent pump. Large pulseless piston pumps cannot be used for this as a continuous closed system is necessary.

Recycling must be stopped before the fastest moving solute overtakes the slowest moving solute, so that the number of cycles possible in any chromatography depends upon the solvent capacity of the closed system and the distribution of the components of the mixture being separated.[392]

Reduced plate height (h_r). In GC and HPLC systems in which the **height equivalent to a theoretical plate** is proportional to the particle size of the column packing, it is possible to make comparisons between the efficiencies of different columns by consideration of the value for the reduced plate height, obtained from the relationship

$$h_r = \frac{h}{d_p}$$

See also **Reduced velocity**.

Reduced velocity (v). A value which, like the **reduced plate height**, is used as an aid to the comparison of different chromatographic columns. It relates the diffusion coefficient for the solute in the mobile phase to the mobile phase velocity and the particle size of the column packing:

$$v = \frac{d_p}{D_M}$$

Reducing factor (j) (Pressure gradient correction factor). *See* **Pressure gradient**.

Refractive index detector. The second most widely used detector in HPLC (the commonest being the **ultraviolet absorption detector**), based upon measurement of changes in the refractive index of the solvent when solutes are eluted from the column. Two main forms of RI detector are in general use: the deflection refractometer (Figure 39*a*) and the Fresnel refractometer (Figure 39*b*).

In the former detector, light is passed through a square cell divided into two chambers by a diagonal transparent partition. Reference solvent is passed through one chamber and solvent from the column through the other chamber. When both chambers contain the same composition solvent, a constant reading is obtained when a beam of light is passed through the cell and reflected back to a position-sensitive photodetector. But if the composition of the column solvent is changed

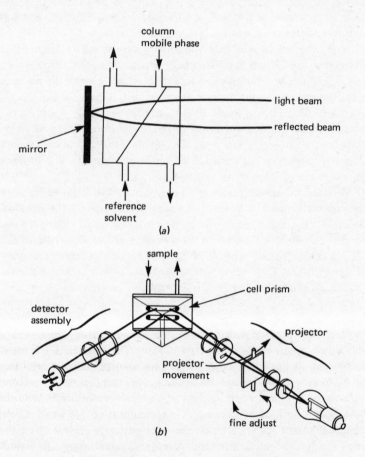

Figure 39. (*a*) Deflection refractometer detector;
(*b*) Fresnel refractometer detector

due to a solute then the altered refractive index causes the beam to be deflected. The magnitude of this deflection is dependent upon the amount of solute in solution; it produces a signal from the position-sensitive photodetector which is amplified and recorded.

In the Fresnel detector both the column solvent and a reference flow of solvent are passed through small cells on the back surface of a prism. Light passed through one wall of the prism is internally reflected from the solvents in the cells on the back wall, to pass out through the third face of the prism to a dual-element photodetector. When the two liquids are identical there is no difference between the two beams reaching the photocell. But when the column effluent containing solute passes through the cell there is a change in the amount of light transmitted to the photocell, and a signal is produced.

Refractive index detectors give both positive and negative traces on the chromatogram as the sign of the signal produced from the photocell depends upon the value of the solute refractive index relative to that for the solvent. Sensitivities of RI detectors are of the order of 5×10^{-7} g cm^3 with cell volumes of 3 μL. A major disadvantage of these detectors is that they are very sensitive to variations in temperature so that control to within ± 0.01 °C is desirable.

Regeneration. Ion exchange resins can be used time and time again if the ions they have taken up in exchange are periodically removed and the resin regenerated to a more useful form. This regeneration is necessary when all the exchangeable sites have become occupied with ions taken from solution. A saturated state such as this occurs, for instance, with water softeners when all the available sodium ions from the resin have been displaced by calcium or magnesium from the water. In the case of domestic water softeners the regeneration is achieved by pouring a highly concentrated sodium chloride solution through the resin in order to displace the undesired ions.

Similarly, the cation exchange resin can also be regenerated to the hydrogen, acid, form if required, by treatment with a molar solution of hydrochloric acid.

Regeneration of anion exchangers follows a similar pattern by employing different solutions of anions that can displace each other.

Relative front. *See* R_F **value.**

Relative retention $(r_{i,s})$. The ratio of the adjusted retention volume (or time) of a substance (i) relative to that for a standard substance (s), its value being less or greater than unity depending upon the peak positions on the chromatogram:

$$r_{i,s} = \frac{V'_{R_i}}{V'_{R_s}} = \frac{t'_{R_i}}{t'_{R_s}}$$

See also **Separation factor.**

Relative retention ratio. *See* **Separation factor.**

Relative sample sensitivity (Lower limit of detection). One of various relationships which have been employed in order to compare sensitivities between different detectors. It is straightforward to estimate as it requires no special equipment or involved calculations. It is determined by injecting decreasing concentrations of the solute into the chromatographic column either until the detection limit is reached, or until there are sufficient points on the response/concentration graph to enable an extrapolation to zero response to be made. This is carried out under defined operating conditions at maximum attentuation. *See also* **Absolute detector sensitivity; Sensitivity; Signal-to-noise ratio.**

Reservoirs. *See* **Solvent reservoirs.**

Resin capacity. *See* **Exchange capacity.**

Resins. A collective word for **ion exchange resins** whether in the form of gels, particles, spheres or membranes, or deposited on solid supports.

Resolution (R_s). The ability of a chromatograph to separate two peaks, expressed in mathematical terms in several ways. The most frequently used equation is

$$R_s = \frac{2(t_{R_2} - t_{R_1})}{w_{b_1} + w_{b_2}}$$

That is, the resolution is equal to twice the distance between the two peak maxima divided by the sum of the peak widths.

An alternative expression, introduced by Purnell[393] and used particularly for HPLC, relates the resolution to the square root of the number of theoretical plates as well as to the **separation factor** α:

$$R_s = \frac{1}{4}\left(\frac{\alpha - 1}{\alpha}\right)\left(\frac{k}{1+k}\right)n^{1/2}$$

Response time. The total response time between a solute passing into the detector cell and the pen starting to move on the chart recorder is determined by three factors.[394] First there is the speed of response of the detection process (the physical or chemical property being measured) used as the basis for the detector. Second there is a detector **dead volume** and third there is the time delay on the electronic equipment. Of the three factors the third is essentially comparable between all detectors and is the smallest contributor to the response time. The most variable of the three is the detector dead volume, which differs greatly, not only between the various detectors but also between detectors of the same type. The response time can be reduced by diminishing the dead volume.

Retardation factor. *See R_F value.*

Retention factor $(R_{x,9})$. Comparisons between substances separated by chromatography are frequently made by measuring their movements relative to a standard or series of standards either by distance, time or volume. The values obtained are called retention factors or retention ratios and can be used as a basis for identification. It has been found that for a given GC column and specified temperature the specific retention volumes for members of a homologous series are related by the equation

$$\log V_g = an + b$$

where a and b are constants and n is the number of carbon atoms in the member of the homologous series (n always > 3). This relationship gives a linear plot if n is plotted against $\log V_g$.

Evans and Smith[395,396] proposed that this relationship should be used in order to express all retention volumes with respect to the straight-chain alkane nonane. Thus for a compound containing x carbon atoms compared to nonane (containing 9 carbon atoms):

$$\log \frac{V_{g(x)}}{V_{g(9)}} = a(x - 9)$$

With this relationship a standard curve can also be plotted for values obtained between nonane and other linear alkanes; the retention volumes of unknown solutes can be referred to this calibration plot in order to ascertain their chain lengths. These retention factors have not been as widely applied as the **Kováts indices**.

Retention index. Any of a number of relationships and equations which have been devised to relate molecular dimensions or structural features to **retention times** obtained on GC columns, to assist in the characterization of unknown substances.[397] Both the **retention factor** introduced by Evans and Smith[398,399] and the **Kováts indices** have been used for this purpose, the latter being rather more widely applied.

Retention time (t_R). The time interval from the point of injection to the appearance of the peak maximum on GC and HPLC. Under standard conditions for partition chromatography the value for t_R can be calculated from the equation

$$t_R = \frac{L}{\overline{u}} \left[1 - \frac{V_s}{V_M} K_D \right]$$

For most calculations in GC the retention time is corrected for the dead volume of the column by deducting the retention time for a nonsorbed species (usually air) to give the adjusted retention time t_R':

$$t_R' = t_R - t_M$$

Retention volume (V_R). The volume of mobile phase required to elute a substance from the column. It is equal to the flow rate (F_c) multiplied by the **retention time**:

$$V_R = F_c t_R$$

It is also given by the relationship

$$V_R = V_M + K_D V_s$$

See also **Specific retention volume**.

Reversed phase systems. Standard liquid–liquid chromatograms are run with a hydrophilic stationary phase and a hydrophobic mobile phase; in the reversed phase system it is the mobile phase which is hydrophilic and the stationary phase which is hydrophobic.[400] This means that the mobile phase (commonly a mixed solvent of water with methanol, ethanol or acetonitrile) is the more polar phase and polar solutes will tend to move with this phase rather than remain with the less polar stationary phase. This approach is of increasing importance in HPLC in which octyl or octadecyl functions are covalently bonded to silica surfaces[401] to create the uniform hydrophobic stationary phase. The retention times of samples are dependent upon the carbon chain length as well as the pore size of the silica support.[402]

Reversed phase systems are used particularly for separating hydrocarbons and related compounds which are sparingly soluble in water. **Gradient elution** procedures involving reversed phase systems have been employed[403] with increasing concentrations of organic solvent in the mobile phase.

In PC, reversed phase systems can be applied by soaking the paper in the appropriate solvent before using an aqueous mixed solvent for the mobile phase. For reversed phase TLC a nonpolar stationary phase of silica gel bonded with octadecyl silane is used in conjunction with a polar (H_2O/CH_3OH) mobile phase.

R_F **value** (Relative front; Retardation factor). A measure of the movement of a solute relative to the solvent front in PC, introduced by Consden, Gordon and Martin.[404,405] Since that time the concept of R_F values has been extended to column procedures and to TLC. It is defined as

$$R_F = \frac{\text{distance moved by solute}}{\text{distance moved by solvent front}}$$

It is a constant for a particular compound under specified conditions of stationary and mobile phases and temperature.[406] (Figure 40). In practice such values tend to be more reliable when determined on paper chromatograms than on thin layer chromatograms. Reproducible R_F values are difficult to obtain in reversed phase TLC and procedures for applying correction factors have been devised.[407,408] As values for R_F are always less than unity they are sometimes multiplied by 100; the resulting values are referred to as hR_F values. *See also* **Chromatographic spectrum**.

Ring-air technique. *See* **Ring-oven technique**.

Ring chromatography. *See* **Circular chromatography**.

Ring-oven technique. Essentially a method of spot analysis carried out on filter paper circles,[409] in which a sample placed at the centre of the circle is selectively eluted by appropriate reagents and solvents to form a ring of 22 mm diameter by evaporation on a hot cylindrical block (the ring oven or Weisz ring oven). Only 1 μL of solution is required for many of the procedures. It is superior to conventional spot analysis as the compounds are much more concentrated in the evaporated ring than they would be in standard spot tests, and as a result smaller quantities can be detected. With the development of the ring oven it could be said that chromatography has almost gone full

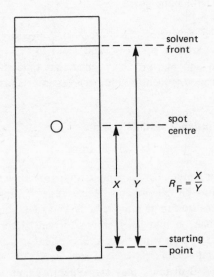

Figure 40. Measurement of R_F value

circle, as it is in many ways an extension of the circular chromatograms (*see* **Circular chromatography**) used by Runge.[410,411]

The ring oven (Figure 41) consists of a cylindrical aluminum block 35 mm high and 55 mm in diameter, with a 22 mm diameter central hole. It is electrically heated to the required temperature and the filter paper circles are held in position by a metal ring. It is suitable for both qualitative and semiquantitative work.[412]

Various modifications of the ring-oven technique exist, including several in which the compounds are concentrated into straight lines by using what is know as the line oven.[413] Of particular interest is the development of the ring-air technique[414] in which a ring of hot air is used to bring about evaporation and concentration. By avoiding direct contact with a metal block it can be used with very volatile solvents and temperature-sensitive materials.

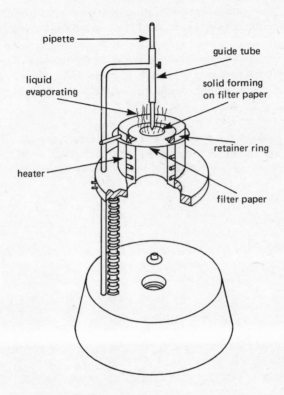

pipette

guide tube

liquid evaporating

solid forming on filter paper

retainer ring

heater

filter paper

Figure 41. Essential features of Weisz ring oven

R_M **value.** A concept introduced to relate R_F **values** of similar compounds to each other. The R_M value is calculated from the R_F value by the equation

$$R_M = \log \left(\frac{1}{R_F} - 1 \right)$$

and is a linear function for series of compounds possessing a regularly repeating group in the molecular structure.[415] It thus gives a straight-line plot for homologous series, although it was originally introduced[416] in studies on flavones and anthocyanins. In any linear presentation it is found that the R_M value increases with the polarity of the compounds. For subsequent members of any series the difference (ΔR_M) between values should be constant. This is the case with homologous series[417] of mainly hydrocarbon chains but less so with aromatic compounds particularly in reversed phase TLC.[418]

Run (Develop). Carry out a chromatographic separation. The various procedures for running a chromatogram are described under **displacement chromatography, elution chromatography** and **frontal analysis**.

S

Sample inlet system. *See* **By-pass injector; Inlet systems; Sample loop; Sample splitter.**

Sample loop (Sample valve). The best means of introducing accurate volumes of liquids and gases on to HPLC and GC columns as it gives the nearest thing to a true **slug injection.** Although many different forms of these devices are manufactured they are all based on the concept of a two-position valve. In the first position (Figure 42) the mobile phase passes continuously through one side of the valve while the sample passes through the other side of the valve. Rotation of the valve through 180° then enables the mobile phase to sweep the fixed volume of sample into the column. Once this has been done the valve can be returned to its original position to be filled with a further sample.

Sample loops are of particular value for repetitive injections and quantitative analyses as no immediate dilution of the sample by the mobile phase occurs. On many sample valves the loop is replaceable or adjustable so that different volumes of sample can be dispensed. *See also* **By-pass injector.**

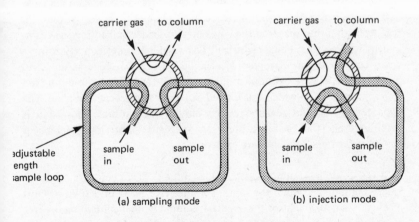

Figure 42. Operation of a sample loop

Sample splitter (Inlet splitter). A device used to apply minute samples to capillary columns, the sample (of μL magnitude) being split into two portions, the smaller of which will be of the nL size suitable for the column. The design of these devices must be such that the sample splitting does not destroy the homogeneity of the sample. Two main types of sample splitter are in common use: calibrated leak devices[419] and dividing valves.[420] Most of these are based upon the principle that the head of the capillary column constitutes only a small proportion of the total outlet from the sample injection block. As the carrier gas enters the block it carries a small amount of the sample into the capillary column while the remainder is vented away through a larger opening. This can be done by having the capillary column held at the axis of a circular hole in the injection block, or by having it forming a narrow arm of a T- or Y-piece at the column head. In either case the amount going into the capillary column can often be further regulated by means of a tapered needle control. *See also* **Effluent splitter**.

Sample valve. *See* **Sample loop.** *See also* **By-pass injector** and **Sample splitter**.

Sandwich technique. A technique whereby admirable thin layer chromatograms can be obtained in the absence of a chromatographic tank. It involves using a second uncoated glass plate over the top of the TLC plate with a thin spacer at the edges on three sides preventing the adsorbent surface from being disturbed. The combination effectively forms a self-contained chromatographic chamber. The lower edge of the sandwich is kept open without a spacer strip in order that the chromatographic solvent in which it is dipped can enter and ascend the plate. The procedure has the advantages that only a small volume of solvent is required and that problems of solvent vapour equilibration which occur in chromatographic tanks are overcome.

SCOT columns. Support-coated open tube (SCOT) columns are capillary columns in which the liquid stationary phase is on a solid support which coats the walls of the capillary column while maintaining the free gas path along the centre of the tube.[421] The result is that there is a greater available stationary phase surface than in **WCOT columns.**

Tubes with internal diameter of 0.5 mm are commonly used for this purpose and have 1000–1200 theoretical plates per metre, so that a 50-metre column has about 50 000–60 000 plates. The increased surface area (compared to WCOT columns) means that sample volumes as great as 0.5 μL can be used without a **sample splitter.**

Sedimentation potential. One of the four electrokinetic phenomena involving movement of the two parts of the electric double layer, considered to be the opposite of **electrophoresis.** It is the electric field formed as a result of charged particles moving relative to a stationary liquid phase. So far this phenomenon has not been applied as a separative or analytical method.

Selective detectors. Detectors which are specially designed to give a signal only when members of particular groups of solutes pass through. In practice nearly all detectors are selective in one way or another, but many detectors are constructed with particular selective applications in mind and they may be of only very limited use outside these areas. Typical of the selective detectors are those intended to detect only compounds containing a particular element or family of elements.[422] The **electron capture detector** is an element-selective detector as it is primarily designed to detect halogenated compounds; the **photometric detector** is another of this group as it has been developed for the detection of phosphorus and sulphur compounds to the extent that its response is 10 000 times greater for these than for hydrocarbons. In HPLC selectivity is rarely intended, although it often arises as a result of the solute having similar properties to the solvent. For instance a **refractive index detector** only gives a reading for substances that lead to refractive index changes when dissolved in the solvent; and an **ultraviolet absorption detector**, which normally operates at a fixed wavelength, is automatically selective for substances that can absorb at the chosen wavelength. *See also* **Thermionic detector.**

Selective elution. A procedure in which a specially chosen eluent is employed that will form nonsorbable compounds with one or more of a limited number of compounds which are to be separated from a complex mixture. When the mixture and the eluent are passed through the column the reacted materials are not retarded and at the same time the passage of unreacted substances is delayed by the column.

Selectivity coefficient ($k_{A/B}$). An equilibrium coefficient quantitatively characterizing the relative ability of an **ion exchanger** to select ion A rather than ion B, and obtained by application of the law of mass action.

Selectivity curve. A curve obtained by plotting the values for K_{av} or the **elution volume** against the logarithm of the molecular weights of solutes on a particular gel column (Figure 43). From such a plot it is possible to pick out a linear region over which the molecular weight range is best suited for gel exclusion on that gel packing. The greater the slope of the selectivity curve the more efficiently does the medium achieve a separation. These curves also serve as the basis for **molecular weight estimation** of macromolecules by gel chromatography.

Sensitivity. The increase in signal intensity per unit increase in solute concentration, that is, dI/dc. The general equation for the calculation of detector sensitivity is

$$\sigma_d = \frac{A_i \sigma_e u_i F_c}{w_i}$$

For GC detectors, sensitivities are frequently expressed in terms of the **Dimbat–Porter–Stross number**, which is a value obtained from a slightly simplified representation of the equation given here.

In HPLC, sensitivity is usually expressed in terms of the **absolute detector sensitivity** or the **relative sample sensitivity** (also called the lower limit of detection).

Separation factor (α) (Relative retention ratio). A term used for the comparison of the net retention times of two consecutive peaks,[423] its value is obtained from any of the following relationships:

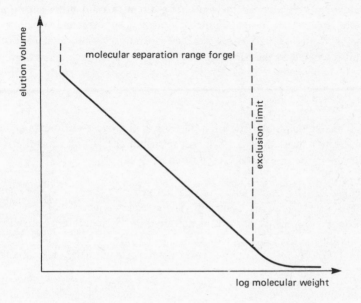

Figure 43. Molecular weight selectivity curve

$$\alpha = \frac{t_{R_2} - t_M}{t_{R_1} - t_M} = \frac{t'_{R_2}}{t'_{R_1}} = \frac{k_2}{k_1}$$

where the subscripts 1 and 2 refer to the two solutes being compared.

The term is used in the Purnell[424] equation for **resolution** and should not be confused with a similar term (α^*) used by Glueckauf.[425] The relationship between the two has been shown[426] to be

$$\alpha^* = \frac{k_2 + 1}{k_1 + 1} = \frac{\alpha k_1 + 1}{k_1 + 1}$$

The values for α and α^* are only identical if column dead space is considered to be negligible compared with the retention volume.

For quantitative analysis to be possible, it is necessary for solutes to be clearly separated on the chromatogram with no overlap occurring between adjacent peaks. Purnell[427] introduced a second separation factor, S, in order to assess whether separation was adequate or not. This is defined by the equation

$$ S = \sigma^2 \left(\frac{\alpha}{\alpha - 1} \right)^2 $$

σ is the standard deviation for the first of the two constituents being considered (assuming a Gaussian distribution). Good quantitative determinations are best achieved when $\sigma^2 > 36$.

Sephadex. The trade name for **dextran gels**, available from the Pharmacia Fine Chemicals Company. These materials consist of fractions of dextran cross-linked with epichlorohydrin.[428] The gels are numbered according to the molecular weight range for which they are best suited (Table 6). The gels in the Sephadex G series are normally only suitable for use with aqueous solutions.

Table 6

Sephadex no.	Molecular weight fractionation range	
	Peptides and proteins	Dextrans
G-10	700	700
G-15	1 500	1 500
G-25	1 000–5 000	100–5 000
G-50	1 500–30 000	500–10 000
G-75	3 000–70 000	1 000–50 000
G-100	4 000–150 000	1 000–100 000
G-150	5 000–400 000	1 000–150 000
G-200	5 000–800 000	1 000–200 000

Much of the early work on **gel chromatography** was carried out on dextran gels.

Sephadex LH20. A modified gel produced by treatment of Sephadex G-25 (*see* **Sephadex**) with aliphatic isocyanates in dimethylsulphoxide. It is stable and swellable in organic solvents.[429] The result of the reaction is the introduction of hydroxypropyl groups into the gel, which increases the ratio of carbon to hydroxyl groups without decreasing the total number of hydroxyl groups, thus producing a combination of lipophilic and hydrophilic properties. Dry beads of Sephadex LH20 have diameters of 25–100 μm.

Septum. Injection of samples by hypodermic syringe for GC and HPLC is commonly made through a self-sealing septum held tightly in the top of the injection port. These septa are made from silicone rubber often reinforced with teflon. Modern septa used for GC are capable of remaining leak-free for more than one hundred injections before they break down and allow carrier gas to escape. For HPLC the septa are made in the form of a sandwich with three layers; the central layer is of a soft silicone rubber which helps to prevent sample blowback, and is reinforced on either side by either teflon or fluorosilicone rubber. These septa are enabled to withstand the high pressures in HPLC by the help of a sliding metal plate which can be displaced to one side to uncover the septum when injections are to be made.

Signal-to-noise ratio. A detector can only be considered to have given a valid signal if there is some measurable response above the level of the normal background noise. Both detector **sensitivity** and the **lower limit of detectability** are dependent upon the level at which the signal can be distinguished. Usually a signal is considered to exist when it is double the noise level, i.e.

$$E_u \geqslant 2R_u$$

where E_u is the minimum amount that can be detected and R_u is the background noise.

Silica aerogels. Rigid **aerogels** which are prepared from porous silica particles and have molecular weight exclusion limits ranging from

6×10^4 to 2×10^6. They are used to form columns with high permeability and have the advantage that no deformation of the matrix occurs in the presence of the solvent or under pressure. They suffer from the disadvantage that it is sometimes necessary to deactivate the silica as its normal adsorption properties can interfere with the true gel exclusion process.

Silica gel. One of the most common materials employed as an adsorbent in chromatography, used extensive in TLC, dry column techniques, GSC and HPLC. Not only is it a good general purpose adsorbent but it is also very cheap. Chemically it is a colloidal form of silica formed by the addition of hydrochloric acid to a concentrated solution of sodium silicate.[430] It possesses a large surface area of about 500 m^2 g^{-1}. For many column procedures it is used with no modification, mesh sizes of about 100–200 being frequently employed. For use in TLC it sometimes has about 5% of gypsum incorporated as a **binder** to give what is known as Silica gel G. To aid detection of the separated spots on TLC plates, it may also have a small amount of a fluorescent indicator incorporated into the powder before being made into the slurry for spreading.

It has been found[431] that silica gel is dissolved by water, strong bases and chloroform when subjected to HPLC conditions and the rate of dissolution is greatly reduced by the use of **bonded stationary phases.**

Silicone oils. A wide variety of silicone oils and gums is available for use as stationary phases in GLC. Their structures are based upon two- and three-dimensional lattices of chains of alternate silicon and oxygen atoms with organic groups, such as methyl, phenyl and vinyl, attached to the silicon atoms. They are semipolar stationary phases used at up to a 10% loading on the support and have operating limits between 200 and 300 °C. One disadvantage with them is that if **bleeding** from the column occurs it can lead to the formation of silicon deposits in the detector. The most widely used of the silicone oils are those sold as SE 30 and MS 550. *See also* **Methyl silicone gums.**

Silylation. Deactivation of support materials and volatilization of organic compounds for GC can both be achieved by procedures known as silylation.

In the former case the support or adsorbent is treated with either **dimethyldichlorosilane** or **hexamethyldisilazane** in order to replace any hydroxyl groups with silyl groups[432] and thus reduce the support polarity.

In the latter case the idea is to prepare volatile silylated derivatives from nonvolatile starting materials.[433] In addition to the two compounds mentioned above, a wide range of silylation reagents has become available.[434] Chlorotrimethylsilane, $(CH_3)_3SiCl$, and N,O-bis-(trimethylsilyl)-acetamide, $CH_3-C[OSi(CH_3)_3]=N-Si(CH_3)_3$, are among those widely used for the purpose of introducing trimethylsilyl (TMS) groups into compounds possessing amino groups[435] as well as hydroxyl groups. Silylation has been of particular value in assisting the GC separation of carbohydrates.[436] Such reactions are straightforward to carry out, simply requiring mixing of the organic compound with the solvent and reagent in the absence of water and warming until fully dissolved.[437] Silylation is just one of the many **derivatization** procedures now employed in chromatography.

Sintered glass plates. One of the problems encountered with home-made TLC is in getting the adsorbent to bind securely to the glass plate. This has been overcome[438] by using sintered glass plates in place of the conventional smooth glass surface.

Slug injection. The ideal method of placing a sample on a chromatograph, involving no immediate dilution of the sample: over a short period of time or volume within the column there is a constant concentration of sample followed by complete mobile phase and no tailing of either. In practice this situation is rarely achieved due to variations in sampling techniques, dilution in the injection block, and the flow of the mobile phase. The nearest approach to slug injection is attained with small volume **sample loops** where diffusion between sample and mobile phase can only occur for a minute period of time before the sample passes from the loop into the column.

Slurry. A thick dispersion of an adsorbent in a liquid, suitable for spreading to form a thin layer on glass plates for TLC or for pouring into a column for adsorption chromatography. In the former case most slurries are made up in water and the wet TLC plate left to dry superficially in the atmosphere before being placed in an oven. In the case of column packings it is common to prepare the slurry in the solvent with which it is intended eventually to run the column. In this case it is necessary to run solvent out from the bottom of the column as the adsorbent settles and more slurry is poured into the top. A slurry is also used for packing the fine columns employed for HPLC. It is especially suitable for packing fine particles which tend to agglomerate if dry-packing techniques are tried, and is also used for packing small particle ion exchange resins which have to be slurried in buffer solutions.

Slurry-packed columns are claimed[439] to be superior to dry-packed columns.

Snake-cage polyelectrodes. *See* **Amphoteric ion exchange resins** and **Ion retardation**.

Soap chromatography. *See* **Ion pairing**.

Solid support. *See* **Support**.

Solubility parameter (δ). A concept introduced[440,441] to enable predictions of solubilities of liquids relative to each other to be made in a semiquantitative manner; in practice they have come to be used as an assessment of solvent polarities. δ is defined as the square root of the cohesive energy density, with units of $(J\ cm^{-3})^{\frac{1}{2}}$, obtained from the relationship

$$\delta = -\left(\frac{E}{V^1}\right)^{1/2}$$

At low vapour pressures this corresponds to the energy of vaporization per cubic centimetre.

Values for δ increase with polarity, ranging from 6.0 for fluorocarbons to 7-8 for aliphatic hydrocarbons, 8-10 for aromatic hydro-carbons, and up to 21 for water.

Some doubt has been cast upon the validity of solubility parameters calculated for polar compounds;[442] despite this they still serve as a useful guide line in drawing up **eluotropic series**.

Solute. Any dissolved component of a mixture placed on a chromato-graphic medium for separative or analytical purposes.

Solute property detectors. Detectors serving to measure physical proper-ties of the solutes as distinct from those of the solvent mobile phase. In this group are the **polarographic radioactivity** and **ultraviolet absorp-tion detectors**. *See also* **Bulk property detectors**.

Solvent migration indices. *See* K_{av} and K_D.

Solvent programming. *See* **Flow programming** and **Gradient elution**.

Solvent regain (S_r). All **xerogels** swell when placed in a solvent, and the porosity of the gel can be established from the swelling characteristics. The solvent regain.[443] is defined as the amount of solvent (in grams) taken up by 1 g of dry xerogel in swelling to an equilibrium state in that solvent. It is determined by allowing a known weight of dry gel to swell to an equilibrium state in excess of the solvent, decanting off the excess solvent, and finally slow centrifuging to remove all the solvent but that which has penetrated the gel. The amount of solvent within the gel is then obtained by reweighing. Solvent regain values are used in calculating the values for the **inner volume** of the gel. *See also* **Water regain**.

Solvent reservoirs. Many chromatographs for HPLC incorporating piston pneumatic pumps (*see* **Pump (piston)** and **Pump (pneumatic)**) employ fixed-volume reservoirs, usually made from stainless steel, that can withstand pressures beyond 6000 pounds per square inch (4×10^7 Nm^{-2}). Capacities of these fixed volume reservoirs range from 250 cm^3 to 1 dm^3. For **gradient elution** two such reservoirs are required.

Solvent series. *See* **Eluotropic series.**

Solvent strength (ϵ^0). A concept introduced by Snyder[444,445] to assist in the selection of suitable solvents for developing adsorption chromatograms on alumina. In this classification all solvents have a value for ϵ^0 within the range of 0.00 for pentane to 1.00 for acetic acid. The series of solvents obtained in this way differs in many respects from other **eluotropic series** and also from the corresponding changes in values of **solubility parameters**, although the general trend of increasing value with polarity is the same.

Sorptiothermal detector. *See* **Thermal adsorption detector.**

Specific capacity. *See* **Exchange capacity.**

Specific retention volume (V_g^θ). The volume of mobile phase per gram of stationary phase corrected to 0 °C (273 K) for defined column conditions for GC. It is related to the other parameters by the following equations:

$$V_g^\theta = \frac{V_N(273\ \text{K})}{w_L T} = \frac{j V_R'(273\ \text{K})}{w_L T} = \frac{j(V_R - V_M)(273\ \text{K})}{w_L T} =$$
$$= \frac{j F_c (t_R - t_M)(273\ \text{K})}{w_L T}$$

Spray reagents. *See* **Chromogenic reagents.**

Spreader. Although plates for TLC can be purchased ready for use it is still very convenient, and often desirable, to prepare the plates in the laboratory as they are required. Most laboratory procedures for spreading plates are based upon that introduced by Stahl.[446]

There are two main parts to the spreading apparatus: the aligning tray and the spreader itself (Figure 44). The purpose of the aligning tray is to ensure that the glass plates are firmly held with their surfaces at a uniform level. The tray is designed to take plates that measure 20 cm across, so that quarter plates (5 cm × 20 cm), half plates (10 cm

X 20 cm) and full plates (20 cm X 20 cm) can all be coated simultaneously. The spreader consists of a metal tube or trough which has a
slit along its length and fits across the aligning tray. It is filled with
slurry and an adjustable gap at the bottom of the spreader allows
the slurry to flow through, forming a thin uniform layer as the spreader
is drawn along the row of glass plates.

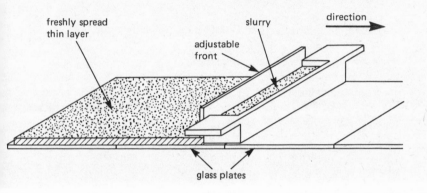

Figure 44. Action of spreader for thin layer plates

Squalane. A liquid stationary phase which is widely used in GLC. Its
correct chemical name is 2,6,10,15,19,23-hexamethyl tetracosane.
It is a nonpolar material used at up to 10% loading on the support, with
an operating limit of 160 °C as its boiling point at 760 mm is 350 °C.
It is particularly suited for separating hydrocarbons and halogenated
hydrocarbons. The related compound squalene[447] possesses double
bonds at the 2, 6, 10, 14, 18 and 22 positions.

Stahl triangle. A triangular relationship introduced by Stahl[448,449] and
used to assist the selection of combinations of stationary phase and
solvent to achieve separation of mixtures by adsorption chromatography,

particularly TLC.

The usual representation of the triangle[450] (Figure 45) has the stationary phase, solvent and mixture characteristics arranged as three equal sectors of a circle. Rotation of the triangle in the centre of the circle with one point directed to the degree of polarity of the mixture gives the general characteristics of the stationary phase and solvent most suitable for the separation at the other two corners of the triangle. Thus a lipophilic mixture requires a nonpolar solvent on an active adsorbent, while a polar mixture is best dealt with in a polar solvent on an inactive adsorbent.

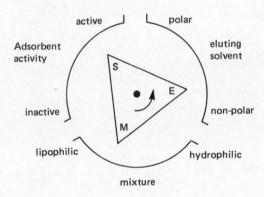

Figure 45. The Stahl triangle

Standard addition. A procedure by which quantitative determinations in GC can be improved in accuracy. To apply the procedure it is necessary to have a pure sample (or a known concentration) of the material for which quantitative measurements are required. The unknown concentration (U) is initially chromatographed to give a peak of area X. Then to the solution of unknown concentration is added a measured amount (B) of the pure compound. As a result the new peak area (Y)

for a chromatographed sample of the same volume is before corresponds to both U and B. Thus the original concentration of U can be obtained from the relationship

$$U = \frac{BX}{Y - X}$$

This relationship only holds if the volume change on addition of B is negligible.

An alternative procedure is to add a constant amount of the unknown concentration to a series of standards of the pure substance and to plot the peak areas obtained against the known concentrations of the original standards. The slope of the line obtained, using the equation below, gives the concentration of the unknown. The line is a calibration plot offset by the amount of the unknown (Figure 46).

$$U = \frac{BZ - CY}{Z - Y}$$

Extrapolation of the plotted line back to the abscissa also gives the unknown concentration (as a negative value). *See also* **Internal standards**.

Start. Of a chromatogram, the point at which a sample is applied and where the process of chromatography commences. In PC and TLC this is a point a short distance above the initial solvent front; in column chromatography it is as near to the column head as the sample can be practically applied.

Stationary liquid volume (V_I) (Intrastitial volume). The volume of the mobile liquid phase which is held stationary in the solid packing or within the pores of a gel. It should not be confused with the **stationary phase volume**.

Stationary phase. A term originally used almost exclusively to refer to the liquid phase held on the solid support in partition chromatography

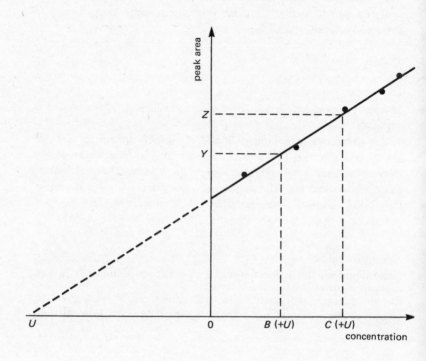

Figure 46. Standard addition plot

systems, but now also used in a wider sense to refer to the immobile phase in all forms of chromatography.

In its original sense the liquids selected as stationary phases for GLC have to meet a number of criteria. They should have a low vapour pressure over a wide temperature range, being nonvolatile and thermally stable at operating temperatures. They should be inert with respect to the mobile phase, the solute, the support and the tube material. They must also be good solvents for a wide range of solutes.

Although more than 400 common substances have been employed as stationary phases, about 90% of GLC separations are carried out on a very small number of liquids.[451] The ten most commonly employed are: Apiezon L, Apiezon M, Silicone SE30, Squalane, Polyethylene glycol 20M, Silicone 550, Silicone QF1, Silicone DC200, Polyethylene glycol 1540, Diethyleneglycol succinate. Preston has recommended[452] that to avoid the proliferation of figures there should be only 20 preferred phases employed for the presentation of retention data. It has also been suggested that satisfactory GLC separations can be achieved in many cases by using just one of four phases: OV-101, V-17, OV-225 and Carbowax 20M, so that an extensive range of packed columns is not necessarily required for good results.

Stationary phase fraction (ϵ_s). The volume of the stationary phase per unit volume of packed column:

$$\epsilon_s = \frac{V_L}{X}$$

Stationary phase volume (V_L). The total volume of the stationary phase liquid on the support material in partition columns. It is used particularly in calculations on **retention volumes** for GLC.

Stepwise elution. In chromatographic procedures employing liquid mobile phases, the use of eluents of different compositions in succession to each other during a single separative run. In most instances the solvents employed are of increasing polarity in order to move the more strongly sorbed substances. For PC and TLC stepwise elution is normally carried out by drying the paper or plate after elution has proceeded a short distance with one solvent before running it with the second solvent. With column procedures the second solvent follows the first down the column. **Gradient elution** is a more refined form of stepwise elution.

Stokes' radius (Molecular radius of gyration). Some high molecular weight materials have been observed to be eluted from **gel chromatography** columns after low molecular weight materials,[453] contrary to normal gel exclusion results. In these cases it has been claimed that the deciding factor is not the molecular weight but the molecular radius of gyration (Stokes' radius). This can occur with molecules and ions which do not act as if they are spherical when passing through viscous media. Under these conditions the Stokes' radius for the molecule is that value for *r* that will satisfy the Stokes' law equation for movement of a sphere through a viscous medium.

Streaming potential. One of the phenomena related to **electrophoresis**, referring to the electric field created when a liquid is made to flow through a stationary capillary of porous material. The potential produced is proportional to the pressure causing the liquid flow and is dependent upon the electric double layer at the solid–liquid interface. Strictly speaking the streaming potential is the potential that must be applied to the system in opposition to that produced by the flow, in order to make the net current flow zero.

Stream splitter. An inclusive term for both **effluent splitters** and **sample splitters.** In the former case the splitter serves to divide the effluent in order that part can pass to the detector and part to the fraction collector or a second detector; in the latter case the sample is split in order to reduce a μL size sample to a nL size for capillary column separations.

Styragel. A trade name for a group of gels based upon styrene and divinylbenzene[454] copolymerized in the ratio 3:1 in toluene[455] and dodecane at 80 °C. The gels have fairly rigid structures and swell only slightly in solvents. They are said to be intermediate between **aerogels** and **xerogels** and can withstand the pressures applied in HPLC.

Subambient operation (Cryogenic operation). **Temperature programming** for GC systems can now cover a range of about 600 °C by the use of cryogenic systems enabling the programming to start at subambient temperatures are low as -180 °C. This type of work is carried out particularly on substances which are very volatile and poorly retained at normal operating temperatures. To achieve the initial subambient temperature the oven is cooled by pumping liquid carbon dioxide or liquid nitrogen through it.[456] The temperature programming is then carried out by progressively decreasing the rate of flow of the liquefied gas until the temperature is above 0 °C, when normal temperature programming can be applied. Systems have been developed which can operate within an accuracy of ± 0.5 °C.

Subambient operation is particularly suitable for separating low molecular weight substances such as the inorganic gases carbon monoxide, carbon dioxide, nitrous oxide and sulphur dioxide.

Support. Traditionally in partition chromatography the solid particles upon which the liquid **stationary phase** is deposited. With the introduction of **WCOT columns** this meaning must now be extended to include the inside surfaces of capillary columns.

Granular support materials should possess: chemical inertness, high surface area per unit volume, low resistance to gas flow by having a high porosity, high mechanical strength and high thermal stability. Supports are usually classified into two main groups[457] of **diatomaceous earths** and nondiatomites. The former group includes **celite**, firebrick and **kieselguhr** while the latter includes carbon, detergents, polymers, glass beads and silica gel.

For GLC columns particle sizes are commonly in the range 60–120 mesh (250–125 μm) while for HPLC they are between 100 and 625 mesh (149–20 μm).

In liquid–liquid partition TLC the solid sorbent for the stationary phase is sometimes known as the carrier rather than as the support.

Support-coated open tube columns. *See* **SCOT columns.**

Surface potential detector. A detector[458] in which a vibrating condenser formed from two dissimilar metals, one of which is gold plated and coated with calcium palmitate, is used as the basis. The surface potential between the plates is modified by any changes in the gas composition and this leads to corresponding changes in the e.m.f., giving a signal proportional to the concentration.[459,460] It can be used to detect 1 part in 10^7 of polar substances and was devised for studying atmospheric pollutants, particularly chlorinated hydrocarbons.

Swelling. Both **ion exchange resins** and **xerogels** swell when immersed in appropriate solvents. In the former case swelling is an osmotic pheno-menon, caused by the solvent entering the resin to dilute the concen-treated ions within it. Swelling may be minimized by using resins with a high degree of cross-linking, but this also reduces the **available capacity** of the resin. The extent of swelling may be expressed in terms of the volume swelling ratio, which is the ratio of the swollen volume to the true dry volume, and by the weight swelling in solvent, which is the gram weight of solvent taken up by 1 g of dry ion exchanger or gel.

In the case of gels the extent of swelling varies with the nature of the solvent and the exclusion limits of the gel; it occurs as a result of the solvent penetrating the pores of the gel, and in this connection it is common to refer to the **solvent regain** value of the gel with respect to the particular solvent. Some gels swell to 5-10 times their initial volume when immersed in the solvent. At room temperature this may take between 1-50 hours but is accelerated by warming the gel and solvent.

Syringes. Precision syringes are employed in both GC and HPLC for manual placing of samples on the chromatographs, and many variations in design exist.

For liquid samples on GC the syringes must be capable of dispensing volumes as small as 0.01 μL and in some cases as large as 500 μL with accuracies of better than 0.1%. Syringes with a total volume of $<$ 1 μL are designed in such a manner that the needle itself serves as the volume barrel, the plunger extending to the tip of the needle so that there is no dead volume (Figure 47).

Syringes for gaseous samples are more conventional in design, having a glass barrel and a plunger with a leak-proof teflon sealing disc. These are designed to measure gas volumes in 0.1 cm^3 divisions with total capacities up to 20 cm^3. Three ranges of sample volumes are employed for GC and HPLC; for capillary columns 0.1 μL or less is required, for analytical columns 0.1-10 μL, and for preparative columns up to 30 cm^3.

High precision GC syringes of the type illustrated can also be used for the injection of small liquid samples on to HPLC at pressures as

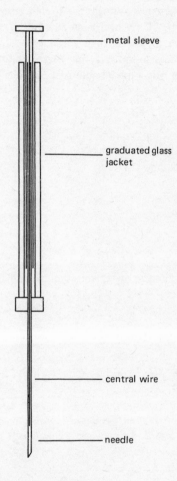

Figure 47. Precision syringe

high as 9000 pounds per square inch (62 MN m^{-2}), but in many cases the specially designed syringes are to be preferred. The main safeguard necessary on HPLC syringes is an adaptor to prevent the plunger being forced backwards by the high column pressure. Because of different injector port design some syringes are made with modified needles, either rounded or completely blunt at the end.

T

Tailing. An effect which is said to have occurred when an asymmetric peak or spot is formed such that the front is sharper or steeper than the rear. The most common causes of this are overloading, particularly in PC and TLC, and simple ionization. *See also* **Fronting**.

Temperature gradient chromatography. *See* **Chromatheromography**.

Temperature programming (Programmed temperature gas chromatography). A technique commonly used in GC in order to increase the rate of elution of highly sorbed species and to decrease the running time for chromatograms. It is used frequently in place of normal ambient operation or isothermal operation at elevated temperatures. During the course of running the chromatogram, the column temperature is increased in a regular manner by means of an electronically controlled heater element in the surrounding column oven. The temperature programme may consist of a linear increase between two fixed temperatures, or a series of steps with intermediate isothermal stages. In some instances temperature programming has been coupled with **subambient operation** so that the programme commences at very low temperatures and rises to very high temperatures with a total temperature range of 400-500 °C. The main result of temperature programming is to reduce the running time of the chromatogram: the more strongly retained compounds are eluted from the column more rapidly as the temperature is raised. This also leads to a reduction in the tailing of the peaks in the chromatogram.[461]

Because of solvent and detector problems, temperature programming cannot usually be applied to HPLC; **flow programming** and **gradient elution** are therefore used instead.

Template. Precise placing of spots on TLC plates without touching or damaging the adsorbent surface is best achieved by means of a protective plastic template. This is designed to cover the surface without touching it, and has a series of small holes regularly spaced along a straight line through which the samples can be applied with **micro-capillaries** or **micropipettes** without the surface being otherwise spoilt.

Theoretical plate. A concept introduced by Martin and Synge[462] in 1941 in order to relate the separation achieved by chromatographic means to the plate theory of distillation processes. They defined the theoretical plate as 'that region in which the average concentration of the solute in the two phases is the same as that which would be obtained if the solute were actually in equilibrium with the two phases in the region'. The physical length of the column to which this relates is called the **height equivalent to a theoretical plate**.

Distribution of a solute determined for a step-by-step equilibration process between a mobile and stationary phase is given by the binomial expansion:

$$(a + b)^n = a^n + na^{n-1}b + \frac{n(n-1)}{2} \, a^{n-2}b^2 \ldots b^n$$

where n is the number of transfers (plates used), a is $1/(k+1)$ and b is $k/(k+1)$. The mass distribution ratio is $k = K(V_s/V_M)$.

The **efficiency of a column** is measured in terms of the number of theoretical plates it possesses along its entire length.

Thermal adsorption detector (Heat of adsorption detector; Sorptio-thermal detector). Any of many devices which employ the temperature changes arising from adsorption/desorption processes as the basis for a universal GC detector. One of the earliest[463] was the use of a thermo-couple embedded in charcoal or in a small amount of the column packing material.

In these devices the temperature of the sensor increases as solutes are adsorbed and correspondingly decreases when they are desorbed. In the more recent devices these temperature changes have been detected with thermistors instead of the thermocouples formerly used. The detector produces a unique first derivative type curve (Figure 48) due to the double process; as a result overlapping peaks can lead to some confusion as competing processes may be taking place simultaneously.

Thermal adsorption detectors have also been used in HPLC[464] and commercial versions are available. In these the thermistor is embedded in silica gel and careful temperature control of the whole detector is necessary.[465]

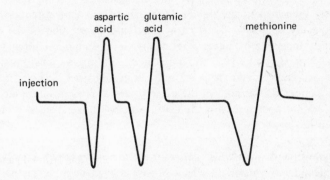

Figure 48. Signal obtained from thermal absorption detector used for an ion exchange chromatogram of amino acids

Thermal conductivity. The rate of transfer of heat (in joules per second) between opposite faces of a 1 m cube when there is a temperature difference of 1 K between the two faces.

In chromatography thermal conductivity is of importance in the **katharometer** as a universal detector in GC. In this type of detector solutes passing with the carrier gas through the detector cell modify the thermal conductivity properties of the gas and lead to temperature changes in the detector filaments or thermistors. Thermal conductivity was first applied in this manner by Claesson.[466,467]

Thermal conductivity detector. *See* **Katharometer.**

Thermal detector. *See* **Thermal adsorption detector.**

Thermionic detector. One of the various modifications of the **flame ionization detector** and the one that has probably been of greatest value. The only real change in the design of the detector is that the hydrogen flame is burnt at a jet with an alkali metal salt tip or with the alkali metal salt as a pellet just above the jet. This leads to a great increase in the sensitivity of the detector to compounds containing nitrogen, phosphorus, sulphur and halogens,[468,469] and it is extensively used in pesticide studies. Caesium and rubidium compounds are frequently employed for the tip. The ionic dissociation of the salt is greatly increased by traces of, for example, organo-phosphorus, enabling levels as low as 10^{-9} to be detected. Although various theories of the operation of the detector have been advanced,[470] none is fully satisfactory. Dimensions in the detector are of critical importance and it does not operate well if the distance between the alkali salt and the tip is too great.

Thermistor. A word which derives from 'thermal resistor', and refers to semiconductor materials which possess a high negative temperature coefficient of resistance. Thermistors were originally developed by Becker *et al*[471] for the Bell Telephone Laboratories; since then the property of decreasing resistance with increasing temperature has been employed in **katharometers** and in **thermal adsorption detectors.**[472]

Thermofractography. A modified sampling technique for TLC in which complex samples, such as plant material, are pyrolysed over a 450 °C temperature range and the volatilized substances spread along the starting line of the TLC plate. The resulting chromatogram is known as a thermofractogram.[473,474]

Thick layer chromatography. *See* **Preparative thin layer chromatography.**

Thin layer chromatography. A type of adsorption chromatography, usually referred to as TLC, which is based upon layers of adsorbent spread on glass plates. Credit for the introduction of TLC is given to Ismailov and Schraiber.[475] In their work dry adsorbents were used, leading to a number of problems due to lack of binding of the surfaces. **Slurries** of alumina to give surfaces 2 mm thick were later tried by Meinhard and Hall,[476] but the technique only really progressed as a result of the extensive study carried out by Stahl and his collaborators.[477,478]

The recommended thickness for the adsorbent in TLC is 150–250 μm, and the uniformity is maintained by use of commercial forms of **spreader** for the adsorbent slurry. TLC is essentially a microanalytical method possessing the virtues of speed, convenience and cheapness. After development of the chromatograms the small amounts of sample applied can be recovered if necessary by extraction from the adsorbent. Most chromatograms are run in the ascending form, and complex mixtures such as sugars or amino acids are best separated by two-dimensional TLC (*see* **Two-dimensional chromatography**).

Commercially produced precoated plates for TLC can be purchased, with highly uniform adsorbent coatings on either glass, aluminium foil or plastic sheets. The main advantage of aluminium foil and plastic sheets, other than convenience, is that they can be cut to any size or shape required, but they also have the disadvantage that they bend in the chromatographic tank if not supported.

Thin layer electrophoresis. *See* **Electrophoresis**.

Thin layer gel chromatography. Gel chromatography, like adsorption chromatography, can be carried out on thin layers but the procedure is slightly different from that for conventional **thin layer chromatography**.[479,480] The gel is allowed to swell in a buffer solution and is applied as a heavy suspension to the glass plate to give a film 1–2 mm thick which must not be allowed to dry out. It is maintained in a closed unit sandwiched between two glass plates with a wick feeder system for the buffer solution, in order to run in a descending manner (Figure 49)[481] at an angle of 10–20°. Solvent flow rates of about 4 cm h^{-1} have been recommended so that the complete chromatogram can take up to 10 h to run.[482] Blue dextran is unsuitable for determining void volumes in this form of gel permeation and **Orange dextran** has been developed for this purpose.

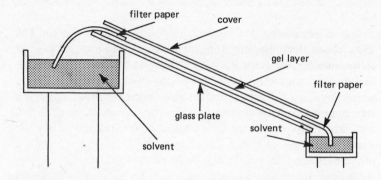

Figure 49. Experimental set-up for thin layer gel chromatography

TLC. *See* **Thin layer chromatography.**

TMS derivatives. *See* **Silylation.**

Total capacity of resins. *See* **Exchange capacity.**

Total volume (V_t). The total volume of a chromatographic column consists of three parts. In partition chromatography these are: the volume of the solid support, the volume of the stationary liquid phase, and the volume of the mobile phase. In adsorption chromatography they are: the volume of the adsorbent, the volume of the mobile phase within the adsorbent matrix, and the volume of the free mobile phase outside the matrix; this classification applies also to ion exchange columns.

In gel chromatography the corresponding three parts are: the volume of the free solvent outside the gel, V_o, the volume of the solvent that has penetrated inside the gel, V_I, and the volume of the solid (aerogel or xerogel) matrix, V_S. The total volume is the sum of these three:

$$V_t = V_o + V_I + V_S$$

and is employed in calculations for the K_D **values** for solutes.

Transport detectors. All transport detectors for HPLC operate on the basic concepts of coating a surface with the eluate, evaporating the solvent, decomposing the solutes, reducing the products and burning the gases at a **flame ionization detector**. For the different methods by which these steps may be carried out, *see* **Belt transport detector; Disc conveyor flame ionization detector; Wire transport detector.**

Traps. Narrow glass U-tubes immersed in solid carbon dioxide or liquid nitrogen. Collection of separated components from GC columns is achieved by passing the eluate through such devices to cause the solutes to condense out. Various forms of traps are employed, and for preparative GC by batch processes several traps will be used fixed to a turntable and programmed to be positioned under the column outlet as the appropriate solute is eluted. *See also* **Fraction collectors.**

Triangulation. A process by which quantitative measurements based upon peak areas can be carried out: tangents are drawn along the peak and the area calculated from the dimensions of the resulting triangle (Figure 50). An alternative procedure, not involving the drawing of tangents, is to multiply the peak height by the width of the peak at half the height (Figure 50). Both methods introduce a small degree of error: in the former case the true area is 1.03 times the triangulated area and in the latter it is 1.06 times the calculated area.[483] *See also* **Disc integrator** and **Electronic integrators.**

Trimethylchlorosilane. *See* **Silylation.**

Tubes. *See* **Capillary columns; Columns; PLOT columns; SCOT columns; WCOT columns.**

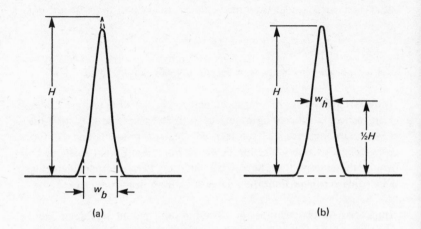

Figure 50. (a) Triangulation; (b) height times width at half-height

Two-dimensional chromatography. A procedure by which improved separation between closely related substances by PC or TLC can be achieved by running the chromatograms in two different solvent systems at right angles to each other. It is applicable to any form of chromatography (or electrophoresis) carried out on a flat surface.

The sample is placed near to one corner of a square sheet of chromatographic paper or square thin layer plate and the chromatogram is run in one direction using the first solvent system (Figure 51). When complete the chromatogram is removed from the chromatographic tank, dried and rotated through 90° in order that the separated row of spots becomes the starting line. The chromatogram is then rerun using the second solvent system. The end result should be a more effective separation over the broader area of the square surface.[484]

Two-dimensional separations are usually easier to carry out by TLC than by PC because of the rigidity of the plate; but in the latter case the lack of rigidity and drying problems have been overcome by the use of special frames which can be used to hold several paper sheets simultaneously.

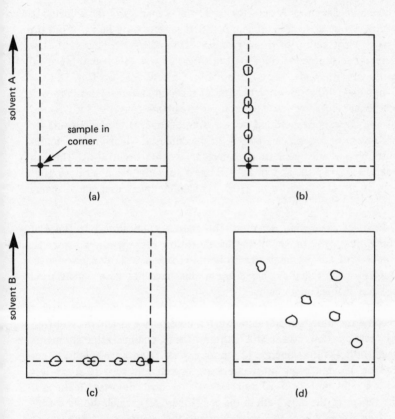

Figure 51. Two-dimensional thin layer chromatography. (a) First development; (b) results of first separation; (c) plate rotated 90° for second development; (d) final chromatogram

U

Ultrasonic detector. A detector based upon changes in the velocity of sound. Its possible development was first suggested by James[485] but the actual construction of the device was achieved by Noble *et al.*[486] The detector measures the frequency of a beat between two opposing ultrasonic vibrations placed at the outlets of a reference gas flow and the column gas flow respectively. The changes in frequency arising when solutes are eluted with the column gas stream are converted into a series of impulses corresponding to the solute concentrations. Helium or hydrogen carrier gases are used[487] in the detector, which gives responses with compounds possessing molecular weights up to 400. The cell volume is 5-50 μL and it gives a linear response from 10^{-2} to 10^{-4} mol at temperatures up to 270 °C. It has the great advantage of being a nondestructive detector.

Ultraviolet absorption detector. The earliest application of ultraviolet absorption was for GC[488] in the detection of polynuclear aromatic compounds for which it was found to have a wide dynamic range. However, GC detectors based upon this have not been widely used because of their unpredictable behaviour.

Since the growth of HPLC the ultraviolet absorption detector has become the most popular detector for liquid systems and many different versions are now manufactured. Those commercial instruments using optical filters operate at one of several fixed wavelengths, either 254 or 280 nm being most popular. The cell volume is 10 μL or less with a light path of 5-10 mm. One of the most important features is the design of the flow path in the cell (Figure 52), which should ensure that diffusion is kept to a minimum and that solutes are cleanly swept through (*see* **Flow cell**). Both single and dual beam instruments are available in a variety of designs. This type of detector is relatively insensitive to temperature and flow changes and can be used with **gradient elution**. The sensitivity is of the order of 5×10^{-10} g cm^{-3}, with the more advanced UV/visible detectors providing full range monochromation between 200-700 nm by means of diffraction gratings so that more selective detection is possible.

196

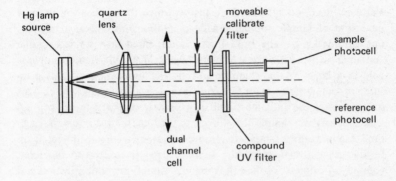

Figure 52. Block diagram of a double-beam UV detector

V

Valveless injection system. A means by which the application of large quantities of sample for batch-wide separations on GC can be carried out automatically. The bulk liquid is dispensed into a fixed-volume container with a flow-back levelling device. The bulk liquid sample is held in a large volume chamber and, by the application of a gas pressure, is forced along a tube to run into a smaller sample vessel. This second vessel can only be filled to a predetermined volume, beyond which any excess sample flows back to the bulk sample reservoir. Each time the measured sample is removed and automatically injected on to the gas chromatograph the sample vessel is refilled to the fixed volume in readiness for the following injection. Control of injection and refilling is coupled with the movement of the **fraction collector.**

van Deemter equation. Probably the most important advance in the development of the theory of chromatography occurred with the publication of what has become known as the van Deemter equation.[489] However, van Deemter and his colleagues considered their work to be an extension of the concepts introduced by Glueckauf (*see* **Glueckauf equation**) for ion exchange and of the earlier theories on partitioning. In its simplest form the equation represents the **height equivalent to a theoretical plate** as being the sum of three terms:

$$h = 2\lambda d_p + \frac{\gamma D_M}{\bar{u}} + \frac{8}{\pi^2} \frac{k}{(1+k)^2} \frac{d_f^2}{D_S} \bar{u}$$

It is often expressed in an abbreviated form as

$$h = A + \frac{B}{\bar{u}} + C\bar{u}$$

The first term in the equation is known as the **eddy diffusion term** and allows for the different paths followed by solutes due to passing around particles of different sizes. The second term is the **molecular diffusion**

term for the contribution due to normal diffusion of molecules taking place while passing along the column. The third term is the **mass transfer term** allowing for the finite rate of transfer between the two phases. In the original presentation of the equation the order of the terms and their names were different.

Various modifications and extensions to the equation have been made;[490,491] the **Golay equation** is a special form of the equation applicable to **capillary columns**. An additional term to the equation has also been introduced[492] in order to make it applicable to preparative gas chromatography.

van der Waals' forces. Adsorption of gases involves a number of different processes, one of the most common of which, called van der Waals' adsorption, occurs to some degree with all substances. It takes place particularly at low temperatures and involves low heats of adsorption of the order of 20–40 kJ mol^{-1}. This physical adsorption occurs as a result of intermolecular forces (the van der Waals' forces) associated with electrostatic attraction between molecules. The success of adsorption chromatography is due to the fact that the adsorption equilibria involving van der Waals' forces are rapidly established and are reversible.

Void volume. *See* **Interstitial volume.**

Volume capacity. *See* **Exchange capacity.**

Volume swelling ratio. *See* **Swelling.**

W

Wall-coated open tube column. *See* **WCOT columns.**

Water regain (w_r). Xerogels employed in gel chromatography swell on addition of solvents, with consequent enlargement of the polymer matrix. Hydrophilic gels which can be used in aqueous solutions swell as a result of water penetrating the gel pores. The extent of the swelling is expressed in terms of the water regain, i.e. the mass of water absorbed by 1 g of the gel. This does not include the water that is trapped between the gel particles which is part of the void volume (*see* **Interstitial volume**).

The water regain is a special form of **solvent regain**.

Water softening. Calcium and magnesium salts present in domestic water supplies impart what is known as hardness to the water, causing the deposit of scale in kettles and boilers. Softening of the water by removal of these salts can be carried out efficiently by ion exchange processes by passing the water through a cation exchange resin in the sodium form, so that the two undesirable metal ions are both replaced by sodium. When the **cation exchanger** in the water softener has become saturated with calcium and magnesium ions it can be regenerated (*see* **Regeneration**) to the sodium form by passing a concentrated sodium chloride solution down the column.

In the domestic water softening process the sulphate and carbonate anions are not normally removed from the water. This can, however, be achieved by using an **anion exchanger** in conjunction with the cation exchanger, either as a separate column or in the form of a **mixed-bed column**. Organic impurities are not normally removed by the water softening process.

Watson–Biemann interface. One of the most widely adopted forms of interfacing between GC and mass spectrometers, developed by Watson and Biemann.[493,494] It is an all-glass, valveless system (Figure 53) in which the pressure reduction and enrichment is effected by means of a sintered glass tube 20 cm long and 8 mm in diameter, terminating in a capillary at each end. The tube acts as a stream splitter, the lighter carrier gas passing preferentially through the sinter while the larger solute molecules continue along the tube. A pressure reduction from 1 atmosphere to 10^{-5} torr is obtained with 50-fold enrichment by using a sinter with holes of approximately 1 μm.

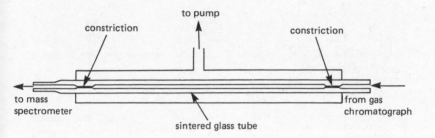

Figure 53. Watson–Biemann separator

WCOT columns. Wall-coated open tube (WCOT) columns were the first types of **capillary column** for GC introduced by Golay,[495,496] with the stationary phase deposited directly on the inside wall of the column. These columns measure from 0.1 to 1.0 mm internal diameter and may be up to 100 m long. Coating the walls is carried out by forcing a 10% solution of the required stationary phase, in an appropriate solvent, through the capillary column under pressure.[497] Sufficient stationary phase is retained on the walls if the solution is forced through at a slow rate of 1–2 cm s^{-1}. **Apiezon L** and high boiling silicone oils have been most frequently employed for these columns. *See also* **PLOT columns** and **SCOT columns**.

Weight swelling in solvent. *See* **Swelling**.

Weisz ring oven. *See* **Ring-oven technique**.

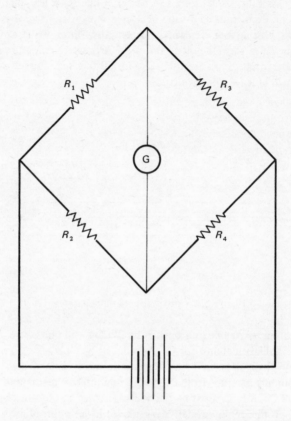

Figure 54. Wheatstone bridge circuit

Wheatstone bridge. Many of the detector systems used in chromato-graphy employ a Wheatstone bridge circuit (Figure 54) in which the detector cell, directly or indirectly, serves as one of the resistance arms. When the bridge is balanced under baseline conditions the four resist-ances in the circuit satisfy the equation

$$\frac{R_1}{R_2} = \frac{R_3}{R_4}$$

and no current flows through the galvanometer (G) or is recorded on the chart recorder. When a solute passes through the detector, leading to a change in the particular detector resistance, it throws the bridge out of balance and a current passes through the meter in proportion to the change in resistance. The balance is restored when no more solute passes through the detector cell.

This type of circuit forms the basis of the operation of the **katharometer** and is used in **electrolytic conductivity detectors**.

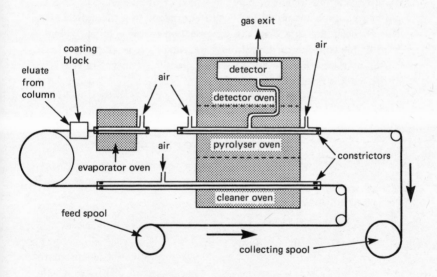

Figure 55. Transport detector

Wire transport detector (Moving wire detector; Phase transformation detector). A form of transport detector[498],[499] for HPLC in which the eluate is deposited upon a fine stainless steel wire that has been cleaned by being heated at 850 °C. The volatile solvent is evaporated and the solute oxidized at 700-800 °C in air to a mixture of carbon dioxide and water. The oxidation products are mixed with hydrogen and passed over a nickel catalyst at 330 °C to give methane, which is burnt in a **flame ionization detector** to give an electrical signal that can be amplified and recorded (Figure 55). The detector gives a linear response over a wide concentration range and is sensitive to 2-3 μg cm^{-3}. The sensitivity has been improved[500] and the noise level reduced by using either a ceramic (kaolin) or ceramic–metal (kaolin and copper) coating to the wire. *See also* **Belt transport detector** and **Disc conveyor flame ionization detector.**

Working capacity of a column. A term which has been defined[501,502] as the maximum weight of pentene which can be injected on to a GC column without increasing the **height equivalent to a theoretical plate**. It is proportional to the area of cross-section of the column. When the limit of the capacity of the column is reached there is a rapid rise in the height of theoretical plates and a corresponding decrease in efficiency as overloading occurs.

X

Xerogels. The most common and most frequently used of the two main classifications of gels used in **gel chromatography**. This group of substances includes all **gels** in which removal of the dispersing agent (the solvent) results in the structure shrinking to an unswollen state (the xerogel). These materials swell to different extents in different solvent systems, the degree of swelling being expressed in terms of **solvent regain**. *See also* **Aerogels**.

Xylenyl phosphate. A GC stationary phase which is useful for separations of aromatic hydroxy compounds such as phenols, but is limited in its applications due to its low maximum operating temperature of 100 °C. It is generally employed at a 5% loading on the support.

Z

Zeolites. Naturally occurring minerals, which have been known for more than one hundred years and over thirty of which have been classified, along with over one hundred synthetic forms.[503,504] They are crystalline hydrated alumino silicates possessing a unit cell formula corresponding to

$$M_{x/n} [(AlO_2)_x (SiO_2)_y] z H_2 O$$

where n is the valency of the metal M (usually sodium, potassium, magnesium or calcium), z is the number of water molecules, and the ratio of $y:x$ is between 1 and 5 depending upon the zeolite structure. The general structure is of a three-dimensional arrangement of AlO_4 and SiO_4 tetrahedra joined together by their shared oxygen atoms. These form a framework of regular pore sizes between 3 and 10Å. They are capable of reversible adsorption-desorption processes,[505] such as hydration–dehydration and ion exchange (*see* **Ion exchanger**) and can sorb molecules with dimensions smaller than the pores of the zeolite structure.[506] Zeolites are the most important of the **molecular sieves** as a result of the very large internal surface area relative to the external surface area.

The zeolites have been extensively studied for their ion exchange properties for cations, although their greatest use is in the petroleum industry for the adsorption of the gases from the cracking processes.

Zone electrophoresis. *See* **Electrophoresis.**

Zones. Individual substances separated on chromatographic columns or TLC plates are sometimes referred to as zones on the chromatogram, although the expression has been mainly used for the visible bands obtained by separating coloured compounds on conventional adsorption and partition columns.

Zwitterions. A number of chemical compounds when in aqueous solution are believed to exist as dual ions carrying a positive charge at one end and a negative charge at the other. These are often called zwitterions or **ampholytes**, of which the most well-known are the amino acids.

REFERENCES

1. H.M.N. Irving, H. Freiser and T.S. West, *Compendium of Analytical Nomenclature, Definitive Rules 1977*, IUPAC, Pergamon Press, Oxford and New York (1978)
2. L.S. Ettre, *J. Chromat.*, **165**, 235 (1979)
3. L.S. Ettre, *J. Chromat.*, **220**, 29 (1981)
4. D.H. Whiffen (ed.), *Manual of Symbols and Terminology for Physicochemical Quantities and Units*, IUPAC, Pergamon Press, Oxford and New York (1979)
5. 'Quantities, Units and Symbols', *A Report by the Symbols Committee of the Royal Society*, London (1975); *Addenda* (1981)
6. M. Tswett, *Proc. Warsaw Soc. Nat. Sci. Biol. Sec.*, **14**, no. 6 (1903)
7. M. Tswett, *Ber. Deut. Bot. Ges.*, **24**, 234, 361 and 384 (1906)
8. D.T. Day, *Science*, **17**, 1007 (1903)
9. L. Reed, *Proc. Chem. Soc.*, **9**, 123 (1893)
10. W. Haller, *Nature*, **206**, 693 (1965)
11. D.H. Campbell, E. Leuscher and L.S. Lerman, *Proc. Nat. Acad. Sci. U.S.A.*, **37**, 575 (1951)
12. P. Cuatrecasas and C.B. Anfinsen, *Methods in Enzymology* (ed. W.B. Jokoby), Academic Press, New York, p. 22 (1971)
13. H. Guilford, *Chem. Soc. Rev.*, **2**, 249 (1973)
14. H.H. Weetall, *Separation and Purification Methods* (eds. E.S. Perry, C.J. Van Oss and E. Grushka), Marcel Dekker, New York, vol. 2, p. 199 (1974)
15. K.W. Williams, *Lab. News*, no. 71, 9 (1974)
16. J. Turková, *Affinity Chromatography*, Elsevier, Amsterdam, London and New York (1978)
17. R.A. Olsson, *J. Chromat.*, **176**, 239 (1979)
18. P. Cresswell, *J. Biol. Chem.*, **254**, 414 (1979)
19. A. Polson, *Biochim. Biophys. Acta.*, **50**, 565 (1961)

209

20. C. Araki, *Bull. Chem. Soc. Japan,* **29**, 543 (1956)
21. S. Hjertén, *Biochim. Biophys. Acta.,* **62**, 445 (1962)
22. B. Russell, T.H. Mead and A. Polson, *Biochim. Biophys. Acta.,* **86**, 169 (1964)
23. S. Hjertén, *Biochim. Biophys. Acta.,* **79**, 393 (1964)
24. C. Araki, *Bull. Chem. Soc. Japan,* **29**, 543 (1956)
25. H. Determann, *Gel Chromatography*, Springer-Verlag, New York, p. 30 (1968)
26. D.H. Spackman, W.H. Stein and S. Moore, *Anal. Chem.,* **30**, 1190 (1958)
27. K.A,. Piez and L. Morris, *Anal. Biochem.,* **1**, 187 (1960)
28. S. Blackburn (ed.), *Amino Acid Determination*, 2nd edn, Marcel Dekker, New York (1978)
29. P.B. Hamilton, *Anal. Chem.,* **32**, 1779 (1960)
30. P.B. Hamilton, *Anal. Chem.,* **35**, 2055 (1963)
31. J.A. Mikes and L.I. Kovacs, *J. Polym. Sci.,* **59**, 209 (1962)
32. M.J. Hatch, J.A. Dillon and H.B. Smith, *Ind. Eng. Chem.,* **49**, 1812 (1957)
33. J.E. Lovelock, *Nature,* **182**, 1663 (1958)
34. J.E. Lovelock, *J. Chromat.,* **1**, 35 (1958)
35. J.E. Lovelock, *Anal. Chem.,* **33**, 162 (1961)
36. S. Prydz and R. Tyvka, *Radiochromatography*, Elsevier, Amsterdam, London and New York (1978)
37. J.J. van Deemter, F.J. Zuiderweg and H. Klinkenberg, *Chem. Eng. Sci.,* **5**, 271 (1956)
38. J.J. van Deemter, F.J. Zuiderweg and H. Klinkenberg, *Chem. Eng. Sci.,* **5**, 271 (1956)
39. J.M. Vergnaud, E. Gregeorges and J. Normand, *Bull. Soc. Chim. France,* 1904 (1964)
40. J.J. van Deemter, F.J. Zuiderweg and H. Klinkenberg, *Chem. Eng. Sci.,* **5**, 271 (1956)
41. P. Chovin, J. Lebbe and H. Moureu, *J. Chromat.,* **6**, 363 (1961)
42. F.A. Gunther, R.C. Blinn and D.E. Ott, *Anal. Chem.,* **34**, 302 (1962)
43. H.W. Johnson, E.E. Seibert and F.H. Stross, *Anal. Chem.,* **40**, 403 (1968)

44. L. De Mourges and V. Rochina, *Bull. Soc. Chim. France*, 729 (1962)
45. R.R. Goodall and A.A. Levi, *Nature*, **158**, 675 (1946)
46. V. Betina, *J. Chromat.*, **78**, 41 (1973)
47. J.N. Miller, *J. Chromat.*, **74**, no. 2, 355 (1972)
48. E. Grushka (ed.), *Bonded Stationary Phases in Chromatography*, Ann Arbor Science, Michigan (1974)
49. C. Horváth and S.R. Lipsky, *Anal. Chem.*, **41**, 1227 (1969)
50. H. Brockmann and H. Schodder, *Ber. Dtsch. Chem. Ges.*, **74B**, 73 (1941)
51. S. Heřmánek, V. Schwarz and Z. Cĕkan, *Collect. Czechoslov. Chem. Commun.*, **26**, 3170 (1961)
52. A.J.P. Martin, *Vapour Phase Chromatography 1956* (ed. D.H. Desty), Butterworth, London, p. 2 (1957)
53. M.J.E. Golay, *Gas Chromatography* (ed. V.J. Coates), Academic Press, New York, p. 1 (1958)
54. M.J.E. Golay, *Gas Chromatography, 1958* (ed. D.H. Desty), Butterworth, London, p. 36 (1958)
55. R. Kaiser, *Gas Phase Chromatography, Vol. 2: Capillary Chromatography* (trans. P.H. Scott), Butterworth, London (1963)
56. M.J. Hartigan, H.D. Hoberecht and J.E. Purcell, *Int. Lab.*, November/December, 30 (1973)
57. K.W. Williams, *Lab. News*, no. 71, 9 (1974)
58. H.H. Weetall, *Nature*, **223**, 959 (1969)
59. H.H. Weetall, *Biochem. J.*, **117**, 257 (1970)
60. L. Rohrschneider and E. Pelster, *J. Chromat.*, **186**, 249 (1979)
61. F. Helfferich, *Ion Exchange*, McGraw-Hill, New York (1962)
62. R. Paterson, *An Introduction to Ion Exchange*, Heyden-Sadtler, London (1970)
63. O. Weigel and E. Steinhoff, *Z. Krist.*, **61**, 125 (1925)
64. F. Helfferich, *Ion Exchange*, McGraw-Hill, New York (1962)
65. A. Sugii, N. Ogawa and H. Hashizume, *Talanta*, **26**, 189 and 970 (1979)
66. N. Chikuma, M. Nakayama, T. Itoh and H. Tanaka, *Talanta*, **27**, 807 (1980)

212   References

67. V. Rezl and J. Janak, *J. Chromat.*, **81**, no. 2 (1973); *Chromat. Rev.*, **17**, no. 1, 233 (1973)
68. S. Pennington and C.E. Meloan, *Anal. Chem.*, **39**, 119 (1967)
69. N.M. Turkel'taub *et al*, *Zh. Fiz. Khim.*, **27**, 1827 (1953)
70. A.A. Zhukhovitskii *et al*, *Dokl. Akad. Nauk. S.S.R.*, **77**, 435 (1951); and **123**, 1037 (1958)
71. A.G. Nerheim, *Anal. Chem.*, **32**, 436 (1960)
72. H.M. Liebich, A. Pickert, U. Stierle and J. Wöll, *J. Chromat.*, **199**, 181 (1980)
73. D.F.G. Pusey, *Chem. Brit.*, **5**, 408 (1969)
74. D.R. Browning (ed.), *Chromatography*, McGraw-Hill, Maidenhead (1969)
75. F.F. Runge, *Zur Farbenchemie*, Mittler u. Sohn, Berlin (1850)
76. F.F. Runge, *Der Bildungstrieb der Stoffe*, Selbstberiag, Oranenburg (1855)
77. G. Hesse and B. Tschachotin, *Naturwiss.*, **30**, 387 (1942)
78. D. Waldi, *Thin Layer Chromatography* (ed. E. Stahl), Academic Press, New York, p. 483 (1965)
79. N.A. Izmailov and M.S. Schraiber, *Furmazia (Sofia)*, **3**, 1 (1938)
80. J.E. Meinhard and N.F. Hall, *Anal. Chem.*, **21**, 185 (1949)
81. E. Stahl, *Perfumerie u. Kosmetik*, **39**, 564 (1958)
82. E. Kawerau, *Chromat. Methods*, **1**, no. 2, 7 (1956)
83. P.E. Barker and D.H. Huntington, *J. Gas Chromat.*, **4**, no. 2, 59 (1966)
84. L. Luft, US Patent, 3,016,107 (1962)
85. S.H. Byrne, *Modern Practice of Liquid Chromatography* (ed. J.J. Kirkland), Wiley-Interscience, New York, p. 95 (1971)
86. E.L. Durrum, *J. Amer. Chem. Soc.*, **73**, 4875 (1951)
87. D.J. Shaw, *Electrophoresis*, Academic Press, London (1969)
88. A. Liberti and G.P. Cartoni, *Gas Chromatography, 1958* (ed. D.H. Desty), Butterworth, London, p. 321 (1958)
89. D.F. Adams, G.A. Jensen, J.P. Steadman, R.K. Koppe and T.J. Robertson, *Anal. Chem.*, **38**, 1094 (1965)
90. D.M. Coulson and L.A. Cavanagh, *Anal. Chem.*, **32**, 1245 (1960)
91. F.W. Williams and M.E. Umstead, *Anal. Chem.*, **40**, 2232 (1968)
92. L.C. Craig and O. Post, *Anal. Chem.*, **21**, 500 (1949)

93. L.R. Snyder, *J. Chromatog. Sci.*, **8**, 692 (1971)
94. L.C. Craig and O. Post, *Anal. Chem.*, **21**, 500 (1949)
95. R.B. Fischer and D.G. Peters, *Quantitative Chemical Analysis*, 3rd edn, W.B. Saunders, Philadelphia, p. 212 (1968)
96. A.J. Pompeo and J.W. Otvos, US Patent, 2,641,710 (1953)
97. P.G. Simmonds and J.E. Lovelock, *Anal. Chem.*, **35**, 1345 (1963)
98. J.E. Lovelock, G.R. Shoemake and A. Zlatkis, *Anal. Chem.*, **36**, 1411 (1964)
99. E.L. Durrum, *J. Amer. Chem. Soc.*, **73**, 4875 (1951)
100. J. Tranchant (ed.), *Practical Manual of Gas Chromatography*, Elsevier, Amsterdam (1969)
101. R.W. Thomas, *Ion Exchange*, Pergamon Press, Oxford (1970)
102. L.G. Saunders, *Lab. Practice*, August (1963)
103. H. Bethke and R.W. Frei, *J. Chromat.*, **91**, 433 (1974)
104. O.S. Privett and M. Blank, *J. Lipid Res.*, **2**, 39 (1961)
105. J.G. Kirchner, *J. Chromat.*, **82**, 101 (1973)
106. J. Drozd, *Chemical Derivatization in Gas Chromatography*, Elsevier, Amsterdam, Oxford and New York (1981)
107. D.R. Knapp, *Handbook of Analytical Derivatization Reactions*, J. Wiley, New York (1979)
108. F.A. Fitzpatrick, *Anal. Chem.*, **48**, 499 (1976)
109. P. Flodin, *J. Chromat.*, **5**, 103 (1961)
110. A.D. Kelmers, *J. Biol. Chem.*, **241**, 3540 (1966)
111. S. Hjeretén and U. Hellman, *J. Chromat.*, **202**, 391 (1980)
112. T.A. Gough and E.A. Walker, *Analyst*, **95**, 1 (1970)
113. R.A. Keller, *J. Chromatog. Sci.*, **11**, 223 (1973)
114. N. Guichard and J. Buzon, in *Practical Manual of Gas Chromatography* (ed. J. Trenchant), Elsevier, Amsterdam, p. 203 (1969)
115. A.T. James, *Vapour Phase Chromatography, 1956* (ed. D.H. Desty), Butterworth, London, p. 129 (1957)
116. S.H. Byrne, in *Modern Practice of Liquid Chromatography* (ed. J.J. Kirkland), Wiley-Interscience, New York, p. 95 (1971)
117. J. Porath and P. Flodin, *Nature,* **183**, 1657 (1959)
118. P. Flodin, Dissertation, Uppsala University, Sweden (1962)
119. L.C. Craig and T.P. King, *J. Amer. Chem. Soc.*, **78**, 4171 (1955)
120. L.C. Craig and W. Konigsberg, *J. Phys. Chem.*, **65**, 166 (1961)

121. J.R. Wilson (ed.), *Demineralization by Electrodialysis*, Butterworth, London (1960)
122. D.M. Ottenstein, *J. Chromatog. Sci.*, 1, no. 4, 11 (1963)
123. D.W. Turner, *Nature*, 181, 1265 (1958)
124. J.D. Winefordner, *Anal. Chem.*, 33, 515 (1961)
125. S.H. Byrne, in *Modern Practice of Liquid Chromatography* (ed. J.J. Kirkland), Wiley-Interscience, New York, p. 95 (1971)
126. H.P. Williams and D. Winefordner, *J. Gas. Chromat.*, 6, 11 (1968)
127. J.F.K. Huber, *J. Chromatog. Sci.*, 7, 172 (1969)
128. M. Dimbat, P.E. Porter and F.H. Stross, *Anal. Chem.*, 28, 290 (1956)
129. J. Bohemen, S.H. Langer, R.H. Perrett and J.H. Purnell, *J. Chem. Soc.*, 2444 (1960)
130. R.H. Perrett and J.H. Purnell, *J. Chromat.*, 7, 455 (1962)
131. J.J. Szakasits and R.E. Robinson, *Anal. Chem.*, 46, 1648 (1974)
132. T. Cotgreave, *Chem. Ind.*, 689 (1966)
133. H. Dubsky, *J. Chromat.*, 79, 383 (1973)
134. B.J. Davis and L. Ornstein, *Ann. N.Y. Acad. Sci.*, 121, [2], 321 and 404 (1964)
135. D.J. Shaw, *Electrophoresis*, Academic Press, London and New York (1969)
136. I.S. Smith, 'Acrylamide Gel Disc Electrophoresis' in *Chromatographic and Electrophoretic Techniques*, (ed. I.S. Smith), 4th edn, vol. 2, 210 (1976)
137. A.J. McCormack, S.C. Tong and W.D. Cooke, *Anal. Chem.*, 37, 1470 (1965)
138. H.A. Moye, *Anal. Chem.*, 39, 1441 (1967)
139. R.S. Braman and A. Dynako, *Anal. Chem.*, 40, 95 (1968)
140. W.C. Baumann and J. Eichhorn, *J. Amer. Chem. Soc.*, 69, 2830 (1947)
141. B. Loev and K.M. Snader, *Chem. Ind.*, 15 (1965)
142. B. Loev and M.M. Goodman, *Chem. Ind.*, 2026 (1967)
143. J.G. Kirchner, *J. Chromat.*, 63, 3 (1971)
144. J.J. van Deemter, F.J. Zuiderweg and H. Klinkenberg, *Chem. Eng. Sci.*, 5, 271 (1956)
145. J.J. van Deemter, F.J. Zuiderweg and H. Klinkenberg, *Chem. Eng.*

Sci., **5**, 271 (1956)

146. J.H. Purnell, *J. Chem. Soc.,* 1268 (1960)
147. R. Rucki, *Talanta,* **27**, 147 (1980)
148. J. Harley and V. Pretorius, *Nature,* **178**, 1244 (1956)
149. H.J. Arnikar, T.S. Rao and K.H. Karmarker, *Ind. J. Chem.,* **5**, 480 (1967)
150. H.J. Arnikar, T.S. Rao and K.H. Karmarkar, *J. Chromat.,* **26**, 30 (1967)
151. D.M. Coulson, *J. Gas Chromat.,* **3**, 134 (1965)
152. D.M. Coulson, *J. Gas Chromat.,* **4**, 285 (1966)
153. D.M. Coulson, *The Detektor,* Micro Tek Instr. Corp., Texas, vol. 1, p. 1 (1968)
154. W.A. Aue and S. Kalpa, *J. Chromatog. Sci.,* **11**, 255 (1973)
155. J.E. Lovelock and S.R. Lipsky, *J. Amer. Chem. Soc.,* **82**, 431 (1960)
156. P. Devaux and G. Guichon, *J. Chromatog. Sci.,* **8**, 502 (1970)
157. E.P. Grimsrud, D.A. Miller, R.G. Stebbins and S.H. Kim, *J. Chromat.,* **197**, 51 (1980)
158. J.E. Lovelock, *Nature,* **185**, 49 (1960)
159. J.F. Ellis and C.W. Forrest, *Anal. Chim. Acta,* **24**, 329 (1961)
160. D.J. Shaw, *Electrophoresis,* Academic Press, London (1969)
161. L. Michaelis, *Biochem. Z.,* **16**, 81 (1909)
162. R.S. Snyder *et al, Separation and Purification Methods* (eds. E.S. Perry, C.J. Van Oss and E. Grushka), Marcel Dekker, New York, vol. 2, p. 259 (1974)
163. D.J. Shaw, *Electrophoresis,* Academic Press, London (1969)
164. K. Macek and Z. Prochazka, in *Paper Chromatography* (eds. I.M. Hais and K. Macek), Academic Press, New York, p. 115 (1963)
165. J.H. Hildebrand and R.L. Scott, *Regular Solutions,* Prentice-Hall, Englewood Cliffs, New Jersey (1962)
166. L.R. Snyder, in *Modern Practice of Liquid Chromatography* (ed. J.J. Kirkland), Wiley, New York, p. 135 (1971)
167. G.W. Atfield and C.J.O.R. Morris, *Biochem. J.,* **81**, 606 (1961)
168. D.J. Shaw, *Electrophoresis,* Academic Press, London (1969)
169. K.O. Pedersen, *Arch. Biochem. Biophys. Suppl.,* **1**, 157 (1962)
170. I.G. McWilliam and R.A. Dewar, in *Gas Chromatography,* 1958

(ed. D.H. Desty), Butterworth, London, p. 142 (1958)
171. A.T. Blades, *J. Chromatog. Sci.*, **11**, 251 (1973)
172. R.P.W. Scott, *Nature*, **176**, 793 (1955)
173. C.F. Cullis, A. Fish, F.R.F. Hardy and E.A. Warwicker, *Chem. Ind.*, 1158 (1961)
174. P.J. Flory and T.G. Fox, *J. Amer. Chem. Soc.*, **73**, 1904 (1951)
175. Z. Grubisic, P. Rempp and H. Benoit, *J. Polym. Sci. B*, **5**, 753 (1967)
176. J.F.K. Huber, *J. Chromatog. Sci.*, **7**, 172 (1969)
177. H. Felton, *J. Chromatog. Sci.*, **7**, 13 (1969)
178. R. Rucki, *Talanta*, **27**, 147 (1980)
179. L. Mázor and J. Takács, *J. Gas Chromat.*, **6**, 58 (1968)
180. S.H. Byrne, in *Modern Practice of Liquid Chromatography* (ed. J.J. Kirkland), Wiley-Interscience, New York, p. 95 (1971)
181. K. Macek, J. Verčerková and J. Stanislavová, *Pharmazie*, **20**, 605 (1965)
182. J.F.K. Huber, *J. Chromatog. Sci.*, **7**, 172 (1969)
183. J.D. Winefordner, H.P. Williams and C.D. Miller, *Anal. Chem.*, **37**, 161 (1965)
184. 'ASTM Proposed Recommended Practice for Gas Chromatography Terms and Relationships', *J. Gas Chromat.*, **6**, 1 (1968)
185. A.J.P. Martin and A.T. James, *Biochem. J.*, **63**, 138 (1956)
186. A.G. Vitenberg, S.K. Pospelova and B.V. Ioffe, *Neftekhimiya*, **12**, no. 4, 623 (1972)
187. J.T. Walsh and D.M. Rosie, *J. Gas Chromat.*, **5**, 232 (1967)
188. J.H. Griffiths, D.H. James and C.S.G. Phillips, *Analyst*, **77**, 897 (1952)
189. A.J.P. Martin and R.L.M. Synge, *Biochem. J.*, **35**, 1358 (1941)
190. A.T. James and A.J.P. Martin, *Biochem. J.*, **50**, 679 (1952)
191. A.T. James and A.J.P. Martin, *Analyst*, **77**, 915 (1952)
192. G. Hesse and B. Tschachotin, *Naturwiss.*, **30**, 387 (1942)
193. E. Cremer and F. Prior, *Öst. Chem. Zig.*, **50**, 161 (1949)
194. L.S. Ettre, *Int. Lab.*, July /August, 8 (1972)
195. L. Condal-Bosch, *J. Chem. Educ.*, **41**, A235 (1964)
196. D. Welti, *Infrared Vapour Spectra*, Heyden, New York (1970)
197. M.D. Erickson, *Appld. Spectroscopy Revs.*, **15**, 261 (1979)

198. M.C. ten Noever de Braun, *J. Chromat.*, **165**, [3] 207 (1979)

199. R. Ryhage, *Anal. Chem.*, **36**, 759 (1964)

200. E. Becker, *Separation of Isotopes*, Newnes, London, p. 360 (1961)

201. S.R. Lipsky, C.G. Horvath and W.J. McMurray, in *Gas Chromatography, 1966: Proc. 6th Int. Sympos. Gas Chromat. Assoc. Techniques, Rome 1966*, Institute of Petroleum, London, p. 299 (1967)

202. J.T. Watson and K. Biemann, *Anal. Chem.*, **36**, 1135 (1964)

203. J.T. Watson and K. Biemann, *Anal. Chem.*, **37**, 844 (1965)

204. H. Determann, *Gel Chromatography*, Springer-Verlag, New York, p. 30 (1968)

205. H. Determann, *Angew. Chem. Int. Edn.*, **3**, 608 (1964)

206. H. Determann, *Gel Chromatography*, Springer-Verlag, New York, p. 30 (1968)

207. A.R. Cooper, J.F. Johnson and R.S. Porter, *Int. Lab.*, May/June, 38 (1973)

208. R.J. Cuetanovic and K.O. Kutschke, in *Vapour Phase Chromatography* (ed. D.H. Desty), Butterworth, London, p. 87 (1957)

209. E. Glueckcauf, *Trans. Farad. Soc.*, **51**, 34 (1955)

210. E. Glueckauf, K.H. Barker and G.P. Kitt, *Disc. Farad. Soc.*, **7**, 199 (1949)

211. J.J. van Deemter, F.J. Zuiderweg and H. Klinkenberg, *Chem. Eng. Sci.*, **5**, 271 (1956)

212. J.J. van Deemter, F.J. Zuiderweg and H. Klinkenberg, *Chem. Eng. Sci.*, **5**, 271 (1956)

213. J.J. van Deemter, F.J. Zuiderweg and H. Klinkenberg, *Chem. Eng. Sci.*, **5**, 271 (1956)

214. M.J.E. Golay, *Nature*, **200**, 776 (1963)

215. L.R. Snyder, in *Chromatographic Reviews* (ed. M. Lederer), Elsevier, New York, vol. 7 (1965)

216. L.R. Snyder, J.W. Dolan and J.R. Grant, *J. Chromat.*, **165**, 3 and 31 (1979)

217. B.L. Glendenning and R.A. Harvey, *J. Forens. Sci.*, **14**, 136 (1969)

218. L.R. Goldbaum, T.J. Domanski and E.L. Schoegel, *J. Forens.*

Sci., **9**, 63 (1964)

219. R.C. Denney, *Drinking and Driving*, Hale, London, p. 105 (1979)

220. C.L. Mendenhall, J. Macgee and E.S. Green, *J. Chromat.,* **190**, 197 (1980)

221. H. Hachenberg and A.P. Schmidt, *Gas Chromatographic Headspace Analysis*, Heyden, London and New York (1977)

222. P.L. Davis, *J. Chromatog. Sci.,* **8**, 423 (1970)

223. J. Drozd and J. Novak, *J. Chromat.,* **165**, 141 (1979)

224. J.J. van Deemter, F.J. Zuiderweg and H. Klinkenberg, *Chem. Eng. Sci.,* **5**, 271 (1956)

225. A.J.P. Martin, *Disc. Farad. Soc.,* **7**, 332 (1949)

226. D. Dinelli, S. Polzzo and M. Taramasso, *J. Chromat.,* **7**, 447 (1962); and US patent 3,187,486 (1965)

227. A. Zlatkis and V. Pretorius, *Preparative Gas Chromatography*, Wiley-Interscience, New York (1971)

228. P.J. Carr, *Proc. Anal. Div. Chem. Soc.,* **12**, 28 (1975)

229. R. Berry, *Nature,* **190**, 1188 (1961)

230. R.T. Parkinson and R.E. Wilson, *J. Chromat.,* **37**, 310 (1968)

231. J. Bohemen, S.H. Langer, R.H. Perrett and J.H. Purnell, *J. Chem. Soc.,* 2444 (1960)

232. J.J. Kirkland (ed.), *Modern Practice of Liquid Chromatography*, Wiley-Interscience, New York (1971)

233. A. Michaelis, D.W. Cornish and R. Vivilecchia, *J. Pharm. Sci.,* **62**, no. 9, 1399 (1973)

234. D.J. Shaw, *Electrophoresis*, Academic Press, London (1969)

235. A.P. Ryle, F. Sanger, L.F. Smith and R. Kitai, *Biochem. J.,* **60**, 541 (1955)

236. C.A. Williams, *Scient. Amer.*, March, 130 (1960)

237. J. Kohn, *Nature,* **180**, 986 (1957)

238. P. Grabar and C.A. Williams, *Immunoelectrophoretic Analysis*, Elsevier, Amsterdam (1964)

239. S. Day, *European Spectroscopy News,* **26**, 31 (1979)

240. J. Lebbe, see p. 54 in reference 100.

241. J.F.K. Huber, *J. Chromatog. Sci.,* **7**, 172 (1969)

242. R.C. Denney, *Drinking and Driving*, Hale, London, p. 105 (1979)

243. R.C. Denney, *Chem. Brit.,* **6**, 533 (1970)

244. H. Puschmann, *Angew. Makromol. Chem.*, **47**, 29 (1975)
245. D. Harvey and D.E. Chalkley, *Fuel*, **34**, 191 (1955)
246. C.A. Pohl and E.L. Johnson, *J. Chromatog. Sci.*, **18**, 442 (1980)
247. E.L. Streatfield, *Chem. Ind.*, 1214 (1953)
248. R. Paterson, *An Introduction to Ion Exchange*, Heyden-Sadtler, London (1970)
249. R.M. Wheaton and R.E. Anderson, *J. Chem. Educ.*, **35**, 59 (1958)
250. R.W. Thomas, *Ion Exchange*, Pergamon Press, Oxford (1970)
251. R.W. Thomas, *Ion Exchange*, Pergamon Press, Oxford (1970)
252. W. Lautsch, G. Manecke and W. Broser, *Z. Naturforsch.*, **8B**, 232 (1953)
253. D.K. Hale, *Chem. Ind.*, 1147 (1955)
254. M. Lederer, V. Moscatelli and C. Padiglione, *J. Chromat.*, **10**, 82 (1963)
255. H.S. Thompson, *J. Roy. Agr. Soc. England*, **11**, 68 (1850)
256. B.A. Adams and E.L. Holmes, *Chem. Ind.*, **54**, 1T (1935)
257. B.A. Adams and E.L. Holmes, *Chem. Ind.*, **54**, 1T (1935)
258. R.M. Wheaton and W.C. Bauman, *Ind. Eng. Chem.*, **45**, 228 (1953)
259. J.E. Lovelock, *Anal. Chem.*, **33**, 162 (1961)
260. P.F. Varadi and K. Ettre, *Anal. Chem.*, **34**, 1417 (1962)
261. A.J.P. Martin and R.L.M. Synge, *Adv. Protein Chem.*, **2**, 31 (1945)
262. R. Gloor and E.L. Johnson, *J. Chromatog. Sci.*, **15**, 413 (1977)
263. J.H. Knox and R.A. Hartwick, *J. Chromat.*, **204**, 3 (1981)
264. J.H. Knox and G.R. Laird, *J. Chromat.*, **122**, 17 (1976)
265. J.A. Mikes and L.I. Kovacs, *J. Polym. Sci.*, **59**, 209 (1962)
266. M.J. Hatch, J.A. Dillon and H.B. Smith, *Ind. Eng. Chem.*, **49**, 1812 (1957)
267. H. Svensson, *Acta Chem. Scand.*, **15**, 325 (1961)
268. D.H. Leaback and C.W. Wrigley, 'Isolectric Focusing of Proteins' in *Chromatographic and Electrophoretic Techniques* (ed. I. Smith), 4th edn, vol. 2, p. 272 (1976)
269. J.R. Sargent and S.G. George, *Methods in Zone Electrophoresis*, 3rd edn, BDH, Poole, p. 206 (1975)
270. F.M. Everaerts, A. Mulder and T.P.E.M. Verheggen, *Int. Lab.*,

Jan/Feb, 43 (1974)

271. J.L. Beckers and F.M. Everaerts, *J. Chromat.*, **68**, 207 (1972)

272. J.L. Beckers, F.M. Everaerts and W.J.M. Houtermans, *J. Chromat.*, **76**, 227 (1973)

273. P. Delmotte, *J. Chromat.*, **165**, 87 (1979)

274. J. Janak, *Coll. Trav. Chim. Tchécosl.*, **19**, 684, 700 and 917 (1954)

275. C. Rouit, in *Vapour Phase Chromatography* (ed. D.H. Desty), Butterworth, London, p. 291 (1957)

276. N.H. Ray, *J. Appl. Chem.*, **4**, no. 21, 82 (1954)

277. T. Johns and A.C. Stapp, *J. Chromatog. Sci.*, **11**, 234 (1973)

278. C.H. Lochmüller, B.M. Gordon, A.E. Lanson and R.J. Mathieu, *J. Chromatog. Sci.*, **15**, 285 (1977)

279. T.C. Laurent and J. Killander, *J. Chromat.*, **14**, 317 (1964)

280. *Gel Chromatography*, Bio-Rad Laboratories, Richmond, California (1971)

281. E.S. Kováts, *Helv. Chim. Acta*, **41**, 1915 (1958)

282. E.S. Kováts, in *Advances in Chromatography* (eds. J.C. Giddings and R.A. Keller), Edward Arnold, London, vol. 1, p. 229 (1965)

283. G. Schomburg and G. Dielmann, *J. Chromatog. Sci.*, **11**, 151 (1973)

284. E. Janos *et al*, *J. Chromat.*, **202**, 122 (1980)

285. K.P. Hupe, *J. Gas Chromat.*, **3**, 12 (1965)

286. D. Vidrine and D. Warren, *Fourier Transform Infrared Spectroscopy*, **2**, 129 (1979)

287. R.E. Lovins, S.R. Ellis, G.D. Tolbert and C.R. McKinney, *Anal. Chem.*, **45**, 1553 (1973)

288. W.J. McFadden, H.L. Schwartz and S. Evans, *J. Chromat.*, **122**, 389 (1975)

289. D.E. Games *et al*, *J. Chromat.*, **203**, 131 (1981)

290. R.G. Christensen, H.S. Hertz, S. Meiselham and E. White, *Anal. Chem.*, **53**, 171 (1981)

291. D.G. Deutsch and E.T. Hertz, *Science*, **170**, 1095 (1970)

292. J.F.K. Huber, *J. Chromatog. Sci.*, **7**, 172 (1969)

293. R.A. Keller, *J Chromatog. Sci.*, **11**, 223 (1973)

294. C.F. Coleman, C.A. Blake and K.B. Brown, *Talanta, 9,* 297 (1962)
295. I.A. Fowlis and R.P.W. Scott, *J. Chromat.,* 11, 1 (1963)
296. T.A. Gough and E.A. Walker, *Analyst, 95,* 1 (1970)
297. R.A. Keller, *J. Chromatog. Sci.,* 11, 223 (1973)
298. L. De Mourges and V. Rochina, *Bull. Soc. Chim. France,* 729 (1962)
299. J.G. Hendrickson, *Anal. Chem.,* 40, 49 (1968)
300. J.G. Hendrickson and J.C. Moore, *J. Polym. Sci. A1, 4,* 167 (1966)
301. A.R. Cooper, J.F. Johnson and R.S. Porter, *Int. Lab.,* May/June, 38 (1973)
302. S.C. Bevan, T.A. Gough and S. Thorburn, *J. Chromat., 42,* 336 (1969)
303. S.C. Bevan, T.A. Gough and S. Thorburn, *J. Chromat., 43,* 192 (1969)
304. H.M.N. Irving, H. Freiser and T.S. West, *Compendium of Analytical Nomenclature, Definitive Rules 1977,* IUPAC, Pergamon Press, Oxford and New York (1978)
305. J.J. van Deemter, F.J. Zuiderweg and H. Klinkenberg, *Chem. Eng. Sci., 5,* 271 (1956)
306. S.W. Mayer and E.R. Tomkins, *J. Amer. Chem. Soc., 69,* 2866 (1947)
307. A.J.P. Martin and R.L.M. Synge, *Biochem. J., 35,* 1358 (1941)
308. W.O. McReynolds, *J. Chromatog. Sci., 8,* 685 (1970)
309. L.S. Ettre, *Chromatographia, 6,* 489 (1973) and 7, 39 (1974)
310. A.J. McCormack, S.C. Tong and W.D. Cooke, *Anal. Chem., 37,* 1470 (1965)
311. S. Jacobs, *Chem. Ind.,* 944 (1955)
312. D.M. Stromquist and A.C. Reents, *Ind. Eng. Chem., 43,* 1065 (1951)
313. R. Kunin and F.X. McGarvey, *Ind. Eng. Chem., 43,* 734 (1951)
314. H.G. Boman, *Nature, 170,* 703 (1952)
315. M. Brenner, A. Niederwieser, G. Pataki and R. Weber, in *Thin-layer Chromatography* (ed. E. Stahl), Academic Press/Springer-Verlag, Berlin and New York, p. 114 (1965)

316. J.J. van Deemter, F.J. Zuiderweg and H. Klinkenberg, *Chem. Eng. Sci.*, **5**, 271 (1956)
317. J.W. McBain, *Kolloid Z* , **40**, 1 (1926)
318. D.W. Breck, *J. Chem. Educ.*, **41**, 678 (1964)
319. P. Andrews, *Biochem. J.*, **91**, 222 (1964)
320. J.T. Gwynne and C. Tanford, *J. Biol. Chem.*, **245**, 3269 (1970)
321. P.E. Barker and S. Al-madfai, in *Proc. 5th Int. Sympos. Adv. Chromat., Las Vegas, 1969*, Preston Technical Abstracts Co., Evanston, Illinois, p. 123; *J. Chromatog. Sci.*, **7**, 425 (1969)
322. G.R. Fitch, M.E. Probert and P.F. Titley, *J. Chem. Soc.*, 4875 (1962)
323. A. Tiselius, *Naturwiss.*, **37**, 25 (1950)
324. J.R. Sargent and S.G. George, *Methods in Zone Electrophoresis*, 3rd edn, BDH, Poole, p. 9 (1975)
325. J.A. Thoma, *Anal. Chem.*, **35**, 214 (1963)
326. R.A. Keller, *J. Chromatog. Sci.*, **11**, 223 (1973)
327. N. Grubhofer and L. Schleith, *Naturwiss.*, **40**, 508 (1953)
328. G. Losse, H. Jeschkeit, G. Fickert and H. Rabe, *Z. Naturforsch.*, **173**, 419 (1962)
329. E. Selegny, N. Thoai and M. Vert, *Compt. Rend. Ser. C*, **262**, no. 2, 189 (1966)
330. W.F. Dudman and C.T. Bishop, *Can. J. Chem.*, **46**, 3079 (1968)
331. J.N. Miller, *J. Chromat.*, **74**, no. 2, 355 (1972)
332. R. Consden, A.H. Gordon and A.J.P. Martin, *Biochem. J.*, **38**, 224 (1944)
333. R. Consden, A.H. Gordon and A.J.P. Martin, *Biochem. J.*, **41**, 590 (1947)
334. F.F. Runge, *Zur Farbenchemie*, Mittler u. Sohn, Berlin (1850)
335. F.F. Runge, *Der Bildungstrieb der Stoffe*, Selbstverlag, Oranenburg (1855)
336. I.M. Hais and K. Macek, *Paper Chromatography*, Academic Press, New York and London (1963)
337. S.V. Heines, *J. Chem. Educ.*, **46**, 315 (1969)
338. C.F. Schönbein, *Verh. Naturwiss. Ges. Basel*, **3**, 249 (1861)
339. F. Goppelsröder, *Capillaranalyse*, Birkhäuser, Basel (1901); and *Anregungen zum Studium der auf Capillaritats – und Adsorp-*

tionserscheinungen Beruhenden Capillaranalyse, Verlag Helbing und Lichtenhahn, Basel (1906)

340. J. Gasparič and J. Churáček, *Laboratory Handbook of Paper and Thin-layer Chromatography*, Horwood, Chichester and Wiley, New York (1978)
341. A.J.P. Martin and R.L.M. Synge, *Biochem. J.*, **35**, 1358 (1941)
342. C.G. Horvath, B.A. Preiss and S.R. Lipsky, *Anal. Chem.*, **39**, 1422 (1967)
343. R.A. Henry, J.A. Schmit and R.C. Williams, *J. Chromatog. Sci.*, **11**, 358 (1973)
344. J.E. Lovelock, *Anal. Chem.*, **33**, 162 (1961)
345. J.G.W. Price, D.C. Fenimore, P.G. Simmonds and A. Zlatkis, *Anal. Chem.*, **40**, 541 (1968)
346. A.N. Freedman, *J. Chromat.*, **190**, 263 (1980)
347. J. Ševčik and S. Krysl, *Chromatographia*, **6**, 375 (1973)
348. N.J. Baker and R.C. Leveson, *Int. Lab.*, **11**, [5], July/August, 65 (1981)
349. W.E. Dale and C.L. Hughes, *J. Gas Chromat.*, **6**, 603 (1968)
350. A.V. Nowak and H.V. Malmstadt, *Anal. Chem.*, **40**, 1108 (1968)
351. M.C. Bowman and M. Beroza, *Anal. Chem.*, **40**, 1448 (1968)
352. G.G. Guilbault, *Anal. Chim. Acta*, **39**, 260 (1967)
353. W.H. King, *Anal. Chem.*, **36**, 1735 (1964)
354. F.W. Karasek and J.M. Tiernay, *J. Chromat.*, **89**, 31 (1974)
355. A.R. von Hippel, *Dielectrics and Waves*, Wiley, New York, p. 262 (1954)
356. J.L. Shohet, *The Plasma State*, Academic Press, London (1971)
357. M.J.E. Golay, in *Gas Chromatography, 1960* (ed. R.P.W. Scott), Butterworth, London, p. 139 (1960)
358. G.C. Goretti, A. Liberti and G. Nota, *J. Chromat.*, **34**, 100 (1968)
359. I. Halász and C. Horvath, *Anal. Chem.*, **35**, 499 (1963)
360. I. Halász and C. Horvath, *Nature*, **197**, 71 (1963)
361. J.G. Koen, J.F.K. Huber, H. Poppe and G.D. Boef, *J. Chromatog. Sci.*, **8**, 192 (1970)
362. H.B. Hanekamp, P. Bos and R.W. Frei, *J. Chromat.*, **186**, 489 (1979)
363. P.C. Joynes and R.S. Maggs, *J. Chromatog. Sci.*, **8**, 427 (1970)

364. S. Hjertén and R. Mosbach, *Anal. Biochem.*, **3**, 109 (1962)
365. E.R. Adlard, in *Vapour Phase Chromatography* (ed. D.H. Desty), Butterworth, London, p. 98 (1957)
366. C.L. Tipton, J.W. Paulis and M.D. Pierson, *J. Chromat.*, **14**, 486 (1964)
367. J.C. Moore, *J. Polym. Sci.*, **A2**, 835 (1964)
368. J.J. Kirkland, *J. Chromatog. Sci.*, **7**, 7 (1969)
369. H.P. Williams and D. Winefordner, *J. Gas Chromat.*, **6**, 11 (1968)
370. A. Zlatkis and V. Pretorius, *Preparative Gas Chromatography*, Wiley-Interscience, New York (1971)
371. J.J. De Stefano and H.C. Beachell, *J. Chromatog. Sci.*, **8**, 434 (1970)
372. L.V. Berry and B. Karger, *Anal. Chem.*, **45**, no. 9 819A (1973)
373. L.V. Berry and B. Karger, *Anal. Chem.*, **45**, no. 9, 819A (1973)
374. M.T. Jackson and R.A. Henry, *Int. Lab.*, November/December, 57 (1974)
375. L.V. Berry and B. Karger, *Anal. Chem.*, **45**, no. 9, 819A (1973)
376. M.T. Jackson and R.A. Henry, *Int. Lab.*, November/December, 57 (1974)
377. C. Horváth and S.R. Lipsky, *Anal. Chem.*, **41**, 1227 (1969)
378. R.S. Lehrle and J.C. Robb, *J. Gas Chromat.*, **5**, 89 (1967)
379. W. Noble, B.B. Wheals and M.J. Whitehouse, *Forensic Sci.*, **3**, 163 (1974)
380. W. Simon, P. Kriemler, J.A. Voellmin and H. Steiner, *J. Gas Chromat.*, **5**, 53 (1967)
381. F.F. Farré-rius and G. Guichon, *Anal. Chem.*, **40**, 998 (1968)
382. W. Simon, P. Kriemler, J.A. Voellmin and H. Steiner, *J. Gas Chromat.*, **5**, 53 (1967)
383. R.L. Levy, D.L. Fanter and C.J. Wolf, *Anal. Chem.*, **44**, 38 (1972)
384. D. Deur-Šiftar, T. Bistacki and T. Tandi, *J. Chromat.*, **24**, 404 (1966)
385. L.C. Mosier, US Patent, 3,078,647 (1963)
386. J.A. Hunt, *Anal. Biochem.*, **23**, 289 (1968)
387. H.P. Williams and D. Winefordner, *J. Gas Chromat.*, **6**, 11 (1968)
388. J.E. Lovelock, *Anal. Chem.*, **33**, 162 (1961)

389. J.H. Dhont and C. De Rooy, *Analyst*, **86**, 74 (1961)
390. A.J.P. Martin, in *Gas Chromatography* (eds. V.J. Coates, H.J. Noebels and I.S. Fageson), Academic Press, New York, p. 237 (1958)
391. K.J. Bombaugh, W.A. Dark and R.F. Levangie, *J. Chromatog. Sci.*, **7**, 42 (1969)
392. K.J. Bombaugh, *J. Chromat.*, **53**, 27 (1970)
393. J.H. Purnell, *J. Chem. Soc.*, 1268 (1960)
394. I.G. McWilliam, *J. Appl. Chem.*, **9**, 379 (1959)
395. M.B. Evans and J.F. Smith, *J. Chromat.*, **5**, 300 (1961)
396. M.B. Evans and J.F. Smith, *J. Chromat.*, **9**, 147 (1962)
397. J. Tranchant, p. 225 in reference 100
398. M.B. Evans and J.F. Smith, *J. Chromat.*, **5**, 300 (1961)
399. M.B. Evans and J.F. Smith, *J. Chromat.*, **9**, 147 (1962)
400. G.A. Howard and A.J.P. Martin, *Biochem. J.*, **46**, 432 (1950)
401. C. Horváth, *Trends in Anal. Chem.*, **1**, 1, 6 (1981)
402. K. Karch, I. Sebastian, I. Halász and H. Engelhardt, *J. Chromat.*, **122**, 3 and 171 (1976)
403. A. Poile, *Perkin-Elmer Analytical News*, no. 10, 8 (1974)
404. R. Consden, A.H. Gordon and A.J.P. Martin, *Biochem. J.*, **38**, 224 (1944)
405. R. Consden, A.H. Gordon and A.J.P. Martin, *Biochem. J.*, **41**, 590 (1947)
406. K.A. Connors, *Anal. Chem.*, **46**, 53 (1974)
407. J.H. Dhont, *J. Chromat.*, **202**, 15 (1980)
408. D.S. Galanos and V.M. Kapoulas. *J. Chromat.*, **13**, 128 (1964)
409. H. Weisz, *Mikrochim. Acta*, 140 (1954)
410. F.F. Runge, *Zur Farbenchemie*, Mittler u. Sohn, Berlin (1850)
411. F.F. Runge, *Der Bildungstrieb der Stoffe*, Selbstverlag, Oranenburg (1855)
412. H. Weisz, *Microanalysis by the Ring-Oven Technique*, 2nd edn, Pergamon Press, Oxford (1970)
413. J.T. Stock, M.A. Fill and L. Dalla Riva, *School Science Rev.*, 369 (1964)
414. R.D. Davies and V. Pretorius, *Talanta*, **26**, 137 (1979)
415. D. French and G.M. Wild, *J. Amer. Chem. Soc.*, **75**, 2612 (1953)

416. E.C. Bate-Smith and R.G. Westall, *Biochim. Biophys. Acta*, **4**, 427 (1950)

417. S. Marcinkiewicz, J. Green and D. McHale, *J. Chromat.*, **10**, 42 (1963)

418. M. Kuchař, V. Rejholec, B. Brůnová and M. Jelínková, *J. Chromat.*, **195**, 329 (1980)

419. L.S. Ettre and W. Averill, *Anal. Chem.*, **33**, 680 (1961)

420. I. Halász and W. Schneider, in *Gas Chromatography, 1962* (ed. M. Van Swaay), Butterworth, London, P. 287 (1962)

421. D.H. Desty, in *Advances in Chromatography* (eds. J.C. Giddings and R.A. Keller), Edward Arnold/Marcel Dekker, London and New York, vol. 1, p. 199 (1965)

422. D.F.S. Natusch and T.M. Thorpe, *Anal. Chem.*, **45**, 1184A (1973)

423. L.S. Ettre, *J. Chromat.*, **198**, 229 (1980)

424. J.H. Purnell, *J. Chem. Soc.*, 1268 (1960)

425. E. Glueckauf, *Trans. Farad. Soc.*, **51**, 34 (1955)

426. S.H. Tang and W.E. Harris, *Anal. Chem.*, **45**, 1977 (1973)

427. J.H. Purnell, *J. Chem. Soc.*, 1268 (1960)

428. P. Flodin, Dissertation, Uppsala University, Sweden (1962)

429. W. Heitz, H. Ullner and H. Höcker, *Makromol. Chem.*, **98**, 42 (1966)

430. E. de Barry Barnett and C.L. Wilson, *Inorganic Chemistry*, Longman, London, p. 291 (1953)

431. J.G. Atwood, G.J. Schmidt and W. Slavin, *J. Chromat.*, **171**, 109 (1979)

432. J. Bohemen, S.H. Langer, R.H. Perrett and J.H. Purnell, *J. Chem. Soc.*, 2444 (1960)

433. S.H. Langer, P. Pantages and I. Wender, *Chem. Ind.*, 1664 (1958)

434. J. Drozd, *Chemical Derivatization in Gas Chromatography*, Elsevier, Amsterdam, Oxford and New York (1981)

435. G. Schneider, S. Janicke and G. Sembdner, *J. Chromat.*, **109**, 409 (1975)

436. C.C. Sweeley, R. Bentley, M. Makita and W.W. Wells, *J. Amer. Chem. Soc.*, **85**, 2497 (1963)

437. A.E. Pierce, *Silylation of Organic Compounds*, Pierce Chemical Company, Chester (1979)

438. T. Okumura and T. Kadono, *J. Chromat.*, **86**, 57 (1973)
439. J.J. Kirkland, *J. Chromatog. Sci.*, **9**, 206 (1971)
440. J.H. Hildebrand and R.L. Scott, *Regular Solutions*, Prentice-Hall, Englewood Cliffs, New Jersey (1962)
441. J.H. Hildebrand and R.L. Scott, *The Solubility of Non-electrolytes*, 3rd edn, Reinhold, New York (1968)
442. H. Burrell, *Interchem. Rev.*, **14**, no. 3, 31 (1955)
443. K.W. Pepper, D. Reichenberg and D.K. Hale, *J. Chem. Soc.*, 3129 (1952)
444. L.R. Snyder, in *Modern Practice of Liquid Chromatography* (ed. J.J. Kirkland), Wiley, New York, p. 135 (1971)
445. L.R. Snyder, *Principles of Adsorption Chromatography*, Marcel Dekker, New York (1968)
446. E. Stahl, *Pharmazie*, **11**, 633 (1956)
447. E.S. Lower, *Speciality Chemicals*, **1**, no. 2, 34 (1981)
448. E. Stahl, *Pharmaz. Rundsch.*, **1**, no. 2, 1 (1959)
449. E. Stahl (ed.), *Thin Layer Chromatography*, Springer-Verlag/Academic Press, Berlin and New York, p. 136 (1965)
450. D.F.G. Pusey, *Chem. Brit.*, 408 (1969)
451. R. Delley and K. Friedrich, *Chromatographia*, **10**, 593 (1977)
452. S.T. Preston, *J. Chromatog. Sci.*, **8**, 684 (1970)
453. L.M. Siegel and K.J. Monty, *Biochim. Biophys. Acta*, **112**, 346 (1966)
454. J.C. Moore, *J. Polym. Sci.*, **A2**, 835 (1964)
455. J.R. Millar, D.G. Smith, W.E. Marr and T.R.E. Kressman, *J. Chem. Soc.*, 304 (1965)
456. J.A. Giannovario, R.J. Gondek and R.L. Grob, *J. Chromat.*, **89**, 1 (1974)
457. D.M. Ottenstein, *J. Chromatog. Sci.*, **1**, no. 4, 11 (1963)
458. G. Phillips, *J. Sci. Instr.*, **28**, 342 (1951)
459. J.H. Griffiths, D.H. James and C.S.G. Phillips, *J. Chem. Soc.*, 344 (1954)
460. J.H. Griffiths and C.S.G. Phillips, *J. Chem. Soc.*, 3446 (1954)
461. L. Mikkelsen, in *Advances in Chromatography* (eds. J.C. Giddings and R.A. Keller), Edward Arnold/Marcel Dekker, London and New York, vol. 2 (1966)

462. A.J.P. Martin and R.L.M. Synge, *Biochem. J.*, **35**, 91 (1941)

463. M. Blumer, *Anal. Chem.*, **32**, 772 (1960)

464. M.N. Monk and R.N. Raval, *J. Chromatog. Sci.*, **7**, 48 (1969)

465. R.P.W. Scott, *J. Chromatog. Sci.*, **11**, 349 (1973)

466. S. Claesson, *Ark. Kemi. Min. Geol.*, **A23**, no. 1, 133 (1946)

467. S. Claesson, *Ark. Kemi. Min. Geol.*, **24**, no. 7 (1946)

468. A. Karman, *Anal. Chem.*, **36**, 1416 (1964)

469. T.H. Mitchell, J.H. Ruzicka and J. Thomson, *J. Chromat.*, **32**, 17 (1968)

470. V.V. Brazhnikov, M.V. Gur'ev and K.I. Sakodynsky, *Chromat. Rev.*, **12**, 1 (1970)

471. J.A. Becker, C.B. Green and G.L. Pearson, *Elec. Eng. Trans.*, **65**, 711 (1946)

472. A.D. Davis and G.A. Howard, *J. Appl. Chem.*, **8**, 183 (1958)

473. E. Stahl, *Z. Anal. Chem.*, **261**, 11 (1972)

474. E. Stahl, *J. Chromat.*, **165**, 59 (1979)

475. N.A. Izmailov and M.S. Schraiber, *Farmazia (Sofia)*, **3**, 1 (1938)

476. J.E. Meinhard and N.F. Hall, *Anal. Chem.*, **21**, 185 (1949)

477. E. Stahl (ed.), *Thin Layer Chromatography*, Springer-Verlag/ Academic Press, Berlin and New York, p. 136 (1965)

478. E. Stahl, *J. Chromat.*, **165**, 59 (1979)

479. B.G. Johansson and L. Rymo, *Acta Chem. Scand.*, **16**, 2067 (1962)

480. B.G. Johansson and L. Rymo, *Acta. Chem. Scand.*, **18**, 217 (1964)

481. C.J.O.R. Morris, *J. Chromat.*, **16**, 167 (1964)

482. K.W. Williams, *Lab. Practice*, **22**, 306 (1973)

483. L. Condal-Bosch, *J. Chem. Educ.*, **41**, A235 (1964)

484. I.M. Hais, *J. Chromat.*, **187**, 466 (1980)

485. A.T. James, in *Vapour Phase Chromatography, 1956* (ed. D.H. Desty), Butterworth, London, p. 129 (1957)

486. F.W. Noble, K. Abel and P.W. Cook, *Anal. Chem.*, **36**, 1421 (1964)

487. K. Abel, *Anal. Chem.*, **38**, 758 (1966)

488. R.A.W. Johnson and A.G. Douglas, *Chem. Ind.*, 154 (1959)

489. J.J. van Deemter, F.J. Zuiderweg and H. Klinkenberg, *Chem. Eng.*

Sci., **5**, 271 (1956)
490. J.C. Giddings, *Anal. Chem.*, **36**, 1483 (1964)
491. W.L. Jones, *Anal. Chem.*, **33**, 829 (1961)
492. J.C. Giddings and G.E. Jensen, *J. Gas Chromat.*, **2**, 290 (1964)
493. J.T. Watson and K. Biemann, *Anal. Chem.*, **36**, 1135 (1964)
494. J.T. Watson and K. Biemann, *Anal. Chem.*, **37**, 844 (1965)
495. M.J.E. Golay, in *Gas Chromatography* (ed. V.J. Coates), Academic Press, New York, p. 1 (1958)
496. M.J.E. Golay, in *Gas Chromatography, 1958* (ed. D.H. Desty), Butterworth, London, p. 36 (1958)
497. R. Kaiser, *Gas Phase Chromatography, Vol. 2: Capillary Chromatography* (trans. P.H. Scott), Butterworth, London (1963)
498. E. Nyström and J. Sjovall, *J. Chromat.*, **17**, 574 (1956)
499. M.H. Pattison, *Int. Lab.*, July/August, 50 (1972)
500. V. Pretorius and J.F.J. van Rensburg, *J. Chromatog. Sci.*, **11**, 355 (1973)
501. E. Bayer, *Abhandl. Deut. Akad. Wiss. Berlin Kl. Chem. Geol. Biol.*, **1**, 8 (1962)
502. E. Bayer, *J. Chem. Educ.*, **41**, 755 (1964)
503. D.W. Breck, *Zeolite Molecular Sieves*, Wiley-Interscience, London (1974)
504. W.M. Meir, *Molecular Sieves*, Society for Chemistry Industry, London (1968)
505. R.M. Barrer, *Endeavour*, **23**, 122 (1964)
506. D.W. Breck, *J. Chem. Educ.*, **41**, 678 (1964)